The Ph.D. Process

The Ph.D. Process

A Student's Guide to Graduate School in the Sciences

Dale F. Bloom
Jonathan D. Karp
Nicholas Cohen

New York   Oxford
OXFORD UNIVERSITY PRESS
1998

Oxford University Press

Oxford   New York
Athens   Auckland   Bangkok   Bogotá   Bombay
Buenos Aires   Calcutta   Cape Town   Dar es Salaam
Delhi   Florence   Hong Kong   Istanbul   Karachi
Kuala Lumpur   Madras   Madrid   Melbourne
Mexico City   Nairobi   Paris   Singapore
Taipei   Tokyo   Toronto   Warsaw

and associated companies in
Berlin   Ibadan

Copyright © 1998 by Dale F. Bloom, Jonathan D. Karp,
and Nicholas Cohen

Published by Oxford University Press, Inc.
198 Madison Avenue, New York, New York 10016

Oxford is a registered trademark of Oxford University Press

Library of Congress Cataloguing-in-Publication Data
Bloom, Dale F., 1949–
The Ph.D. process : a student's guide
to graduate school in the sciences /
Dale F. Bloom, Jonathan D. Karp, and Nicholas Cohen.
p. cm.   Includes bibliographical references and index.
ISBN 0-19-511889-8 (cloth).— ISBN 0-19-511900-2 (pbk.)
1. Science—Study and teaching (Graduate)
2. Biology—Study and teaching (Graduate)
3. Doctor of philosophy degree.
4. Graduate students.
I. Karp, Jonathan D., 1964–   .
II. Cohen, Nicholas, 1938–   .   III. Title.
Q181.B5574   1998   507'.1'173—dc21   97-37159

3 5 7 9 8 6 4 2
Printed in the United States of America
on acid free paper

Contents

Acknowledgments, ix

Preface, xi

1 Deciding to Go to Graduate School, 3

2 Selecting an Advisor: Whose Lab Is Right for Me?, 20

3 The Stages of Graduate School, 35

4 Classes, Journal Clubs, Lab Meetings, and Seminars, 41

5 The Absent Professor, 50

6 How You Learn, 55

7 Deciding on Research Projects for Your Dissertation, 70

8 Networking, 83

9 Picking a Dissertation Committee, and Defending the Proposal at the Preliminary Oral Exam, 90

10 The Life of a Graduate Student, 99

11 Some Additional Aspects of Graduate School Life: Lab Notebooks, Etiquette, Competition, Luck, 119

12 Do I Belong Here?: Insecurity and Stress, 132

13 Foreign Students: Unique Problems and Stresses, 141

14 On the Art of Scientific Writing, 150

15 What Should Your Goals Be While in Graduate School?, 160

16 Times They Are A-Changing, 175

17 The End Is in Sight: Writing the Dissertation, 186

18 The Final Oral Exam (The Defense), 192

References and Additional Reading, 201

Index, 207

Acknowledgments

We are indebted to the many scientists and graduate students, too numerous to mention individually, who contributed to this book by agreeing to be interviewed or by responding to our questionnaire. Their personal accounts and thoughtful advice on all aspects of the graduate school experience truly made the book come alive. Particular thanks go to: Cynthia Carey, Snezhana Dimitrova, Barry Goldstein, Sheila Kelley, Basia Kruszewska, Julie Rehm, Elizabeth Stein, and S. ThyagaRajan, for their time and astute comments; Ted Melnechuk for providing a wealth of invaluable information; Anne Myers for her insightful and wonderful words; our agent, Richard Balkin, for believing in the project; our editor, Kirk Jensen for perspicacious observations and suggestions; and Gary Boehm, Tracy Callahan, Ineke Cohen, John Frelinger, Paula Geiselman, George Kimmich, Greg Maniero, and Rick Willis for carefully reviewing the manuscript and making perceptive and valuable recommendations.

D.F.B., J.D.K., & N.C.

I am indebted to wonderful scientists who made my experiences as a graduate student at UCLA so meaningful and memorable: Art Arnold, who by example and instruction first taught me how to think like a scientist; George Bloch, who as a postdoc gave so much of his time and was a true mentor; and Roger Gorski, who wisely guided me through the intricacies of systems neuroscience and gave me unusual freedom to explore. Emilio Decima, Carlos Grijalva, Wendy Macklin, Irv Maltzman, Paul Micevych, Charles Sawyer, Elizabeth Stein, and David Whitmoyer are talented and spirited researchers who inspired me and provided assistance and counsel. Fellow graduate students, so much a part of the educational experience, delighted me with their inquiring minds and enthusiasm. It was a joy to be in the company of Laura Allan,

Gayle Boyd, Hao-Dong Cai, Robin Dodson, Julian Levy, Fredricka Martin, Greta Mathews, Andrea Moskowitz, Rich Nahin, Laura O'Farrell, Yehuda Shavit, and June Stapleton. For years of encouragement and support, I wish to recognize with appreciation and love my family: Betty, Richard, Steven, and Ronnie Bloom, Harry Mann, and Maisie Samuel. Finally, I acknowledge my dear friend Thea Iberall, who at the moment of initial indecision, e-mailed those crucial and deciding words: "Go for it!"

D.F.B.

Heartfelt thanks to my mentors: Gerhard Fankhauser and John Boner who introduced me to research; Hans Holtfreter who taught me to ask questions of Nature; Bill Hildemann, a role model whose life remains intertwined with mine; and my past and present graduate students and postdocs who now understand what it's all about.

N.C.

The present form of this book would not have been possible if it were not for the "graduate school underground": the collective students, postdocs, and faculty members who agreed to speak openly. I am privileged to have participated in their catharsis. With them, I proclaim that this book is for the graduate students of the future, from the graduate students of the past.

J.D.K.

Preface

The world of a graduate student in the sciences is privileged and wondrous. One is surrounded daily by skillful researchers bustling about confidently doing science—people who are making genuine strides in the understanding of nature, discovering the knowledge that will fill the textbooks of the future. Some of these researchers may be world leaders in their fields; others may be relatively unknown, but on the verge of great discoveries. In this high-tech environment, at the very forefront of knowledge, life is exciting, challenging, and stimulating. Every day can bring new insights, new directions, new revelations.

And here you are, a new graduate student, about to become an important participant in this restricted and extraordinary world. Surely you are excited and thrilled by the opportunity granted you. But you are likely also to be feeling somewhat confused and perhaps anxious at the prospect. Exactly where do you fit in? What is expected of you? What will your daily life be like? What is the best way for you to proceed to insure the successful completion of your education, and future success in your chosen profession?

Your confusion about these issues is understandable. In fact, it may even persist for quite a while after you arrive. Some students never figure it all out, and their careers suffer or never begin. The confusion and uncertainty stem from a peculiar characteristic of graduate school: *the rules of the game are rarely talked about!*

Contrary to what you might be expecting, unless you have a wonderful mentor, you are unlikely to receive explicit guidance as to what you should or should not be doing in grad school. A graduate program has a certain vague and open-ended character. There is no rigid agenda, no circumscribed plan of action, no set timetable. Little is mapped out for you. You will not take any specific courses on how to be a scientist, and no one will stand by you and teach you all the tricks. How, then, do you learn? Yes, there are some classes that you will be required to take, but will these be your main source of information, and are grades really important? What about research, that singularly most important aspect of graduate school: In whose lab will you work? What will you work on? How do you gain expertise? Who can help you? How hard do you have to work?

Logical reasoning, critical analysis, scientific writing, public speaking, networking, and other such skills are also elements to be mastered during graduate school, but how are they perfected? There are so many components to a graduate education, but which ones are critical for future success, and deserve most of your energy and time? What decisions can make or break your career? Keep in mind that graduate school is not an experiential extension of undergraduate education, where the passing of a sufficient number of courses guarantees one a degree. Nor is it medical school or law school, where a delineated and set curriculum gives order, structure, and time limits to the educational experience. This is graduate school in the sciences, where each student in the program has a different experience, learns different skills and different information, and finishes at a different time; where poor choices can delay the degree by years; where you are relatively free to use your time as you see fit; and where you will sink or swim, depending upon *your own interpretation* of how the system works. It can be startling. Self-discipline and self-initiative rule— you are pretty much on your own, as you will soon find out.

The purpose of this book is to provide you with some insight into this novel system, insight that will help you succeed and make the most of your graduate years. The contents are what a "best friend" would tell you about graduate school and how it should be approached. We describe and explain the "rules of survival and success," for although generally unspoken and cloaked, "rules" they surely are; we believe that many of those who eventually drop out of a graduate program or fail to succeed, just never "got it."

Also described is the daily experience itself: the research, classes, seminars, journal clubs, lab meetings, interactions with peers and professors, manuscript preparation, qualifying exams, professional meetings, oral exams, dissertation preparation, and other elements that comprise this necessarily compulsive existence. Anxiety, frustration, and joy—all normal responses to a grad student's life—are also examined. So, too, are the special problems of foreign students who are strangers to our culture and educational system.

An important part of this book is the inclusion of quotes about graduate school (obtained via questionnaire or personal interview) from numerous people across the country in a diversity of scientific fields, who have been through the process or are now going through the process—professors, scientists in industry, postdoctoral fellows, and current graduate students. These people describe their own experiences and emotions, and offer advice on the variety of challenges and circumstances that constitute graduate school in the sciences. (Unless otherwise indicated, the university listed under each quote represents the graduate school attended, and not the institution in which the respondents now work. The comments do not reflect upon the university or department denoted, as they are the views solely of individuals, and may not be representative; furthermore, some of the respondents graduated a number of years ago, and departments do evolve.)

This is not a book about science (although some helpful hints concerning particular scientific endeavors are given). If anything, it is a book on the "sociology" of academia, and on graduate education in the sciences in particular. Most of the issues we discuss are timeless, and apply to students of all the sciences, biological and physical. Beginning students have many misconceptions about graduate school and what it takes to succeed there. Once you know how the game is played, once you realize what your advisor is really asking from you, once you are familiar with what your life will be like for the next four to seven years, your graduate experience should be predictable, surmountable, and for the most part, enjoyable! We suggest that you read this book in its *entirety* at an early stage in your career planning, so that you can take steps in advance and make wise decisions as choices arise.

The making of a scientist is a slow and sometimes painful process, and how it happens is neither straightforward nor easily described. But it does happen, over many years, from the subtly combined influence of formal discourse, informal chats, observation, successful and unsuccessful experimentation, error, criticism, and praise. The result of this education is a person who thinks deeply, logically, and especially critically, who is knowledgeable in the area in which he or she works, who knows how to correctly design experiments to test his or her hypotheses, and who is technically proficient enough to carry these experiments out—in short, a scientist.

A Note on Organization

This book is organized in a largely chronological order, although beginning students may have some difficulty in seeing this. Opening chapters (Chapters 1 and 2) discuss the decision to go to graduate school and the application procedure, as well as the task of choosing a research advisor. The latter topic is necessary to bring up early, as advisors are sometimes selected *before* a student arrives at the university, or shortly thereafter. These chapters are followed by an overview of the stages of graduate school (Chapter 3), so that the reader can get a general feel for the framework of the educational process, and is thus prepared for the more detailed discussions that follow. Chapter 4 delves into activities that students participate in early on: classes, journal clubs, seminars, etc. As many students also start working in a lab shortly after arrival on campus, we felt it important that they realize beforehand that they are unlikely to be working side-by-side with their research advisor, and we explain why this (sometimes resented) state of affairs is so (Chapter 5). Because their advisors are typically absent from the lab, students may wonder how it is that they ever learn to be scientists; the interesting answers to this complex question are enumerated in Chapter 6. The tricky business of choosing specific research projects for the dissertation, the pressures to specialize, and the necessity for interacting with other scientists and making contacts in the academic community (networking) are issues that the developing scientist must confront (Chapters 7 and 8); ensuing responsibilities, including choosing a dissertation committee and preparing for the preliminary oral exam (Chapter 9), mark the student's entrance into focused research. Following this general account, Chapter 10 focuses on the nature of a graduate student's daily life, with much of the emphasis placed on the trials and tribulations of the student's research years. Elements that are part of that life, such as note-taking, lab etiquette, competition, luck, stress, and the challenges of scientific writing, are discussed in Chapters 11 through 14. Chapters 15 and 16 consider the should-be goals of a student as he/she progresses through the program. Discussed here are objectives to be accomplished *before* graduation that will help ensure career success during these times of limited employment opportunity in science. Finally, the end stages of the graduate process, writing the dissertation and passing the final oral exam, are covered in Chapters 17 and 18.

The Ph.D. Process

1

Deciding to Go to Graduate School

Some of you think that deciding to go on to graduate school is the easiest decision in the world. And it is. You love learning about the natural world, delight in the thought of the investigative process, and are enticed by the idea of contributing to society and its body of knowledge. You have a calling, and proceeding on to advance your education is an effortless move that is undertaken with relish. For you, deciding *not* to go on to graduate school is the difficult decision to make. You probably have heard through the grapevine that there is little employment opportunity, money, or job security in science today. Funding for research is tight, and hard work and dedication no longer guarantee a career. If you are convinced that there is nothing you would rather do than dedicate your life to science, then these (very true) observations probably should not dissuade you; do not give up on your dream. Likewise, if you love science, and strongly desire a career in a science-*related* field (such as wildlife management) where a graduate degree would be helpful, the trials you are about to encounter should be bearable. However, if your interest is piqued by science, but you are *unsure* if you want to make it a career, graduate school may not be for you. Without a *powerful* drive, the full-time effort and unflagging perseverance required, combined with the gloomy employment situation, will make you miserable. Graduate school is a long, hard road, with inherent major stresses, and, although good times and enchanted moments will certainly transpire, the many years that you will spend there will be no picnic. You will work very hard, be tested and challenged repeatedly, forfeit much, and feel overwhelmed much of the time.

♦   Be certain that graduate school is something that you really
want to do. It seems like some people apply to graduate school because they don't know what else to do at that point in time. Once
you get there, you will work harder than you ever have in your life
because you need to balance lab research, courses, seminars, etc.,
with sleeping and eating. You will be expected to do more than
you have time for, and unless a Ph.D. is something you REALLY
want, you probably will not succeed. I have figured this out the
hard way! (Graduate student, Biology, University of Virginia)

Unless you truly love science, *and* are willing to put up with
frustration, arduous hours, and poor job prospects, graduate school
will prove to be a very difficult venture.

Which Degree?

That said, a student determined to pursue a graduate degree must
first decide if a Ph.D. is necessary, or if a master's degree will suffice. A master's degree, which takes one to three years to obtain,
prepares one for technical positions in academia or industry, high-
school teaching, and a variety of other science-related professions,
such as science journalism. A master's program can also function
as a testing ground; it is an excellent means of determining if one
enjoys research enough to eventually pursue a doctorate. A Ph.D.
degree, which typically takes four to seven years to obtain, prepares one for teaching and research at the college and university
level, and for certain nonacademic careers (see Chapter 16), such as
industrial research. Admission standards for the master's are less
demanding than those for the Ph.D. and the master's program—
which requires coursework (often the same as that for the Ph.D.)
and either a modest (in scope) research project and thesis, or a
library-researched paper—is shorter; a comprehensive exam (shorter
than the one for the Ph.D.) and/or a final oral exam may or may
not be required. In essence, a master's degree program in many departments may be a scaled-down version of a Ph.D. program. However, master's students typically pay their own way and do not
customarily receive financial support for tuition or living expenses from their departments or advisors.

Students who are interested in pursuing a career in medicine,
in addition to one in research, may want to consider applying
to those universities that offer a joint M.D./Ph.D. degree. These

highly selective programs enable students to complete both degrees in a minimum amount of time. Students take medical and graduate school classes, perform years of research, defend a dissertation, and carry out clinical rotations.

Prior Research Experience

It is not a good idea to go into graduate school "cold." Prior research experience is extremely helpful for making the adjustment to lab or field work easier, and is a *selling point for admission.* Having research experience reveals that you are knowledgeable about the research process and, since you are applying to graduate school, obviously enjoy participating in it. In fact, since most entering graduate students do indeed arrive with a certain amount of research experience, some such background is expected. You will be starting with a handicap if you have never performed experiments before. Volunteer to work on a professor's project at your undergraduate institution. Work for a year or two as a lab technician—many people take a job as a technician in academia or industry to gain competence and test the waters before going on to graduate school; those who have held a job before pursuing the Ph.D. tend to be more mature, focused, and successful graduate students.

♦ The biggest mistake people make is deciding to go to grad school because they don't know what else to do, or because their parents expect them to, or because it is "the natural next step." Graduate school is no party; it is challenging intellectually and emotionally. All grad students go through a period of self-discovery and it can be very painful unless you enjoy what you are doing. Someone shouldn't go to grad school because they have their heart set on obtaining a specific position when they are done, or because they don't know what else to do; they should go because they want to accept the ultimate challenge!

I went to grad school because I knew I didn't want to be a technician. I knew that in order to achieve the level of independence that I wanted in a job, I needed a Ph.D. Without my industrial experience as a technician, grad school would have been a lot more traumatic. Because I had previous experience in a lab, I was not as intimidated by lab work, instrumentation, etc. As a result I had an enjoyable grad school experience. (Ph.D., Chemistry, University of Rochester)

♦   I think I knew what I was getting into. I had been an undergraduate at an institution that had a serious Ph.D. program yet
was small enough that good undergraduates could get lots of personal attention from the faculty, and research positions in their
laboratories. I spent my junior and senior years working in two
different research groups in almost the same style as a graduate
student. I had a pretty good idea of what graduate student life was
like and what research toward a Ph.D. degree involved. (Ph.D.,
Chemistry, University of California, Berkeley)

Selecting a School

There are two basic ways to select a school for graduate work. One
way to make your decision to attend University X is based on your
interest in a particular professor who teaches there. If you are
eager to work under a specific faculty member because you are familiar with and excited about his or her work, you should contact
that individual directly (you can access information about the research interests of professors by going to the university's web page
on the internet). Write a letter explaining your interest, describe
your background, and inquire whether you have a realistic chance
of being accepted into his or her lab. If the professor indeed finds
you an attractive candidate, he/she might mention this mutual regard to the department's admission committee (a group of faculty
members), who may then give special consideration to your application (especially if the professor is willing to obligate his or her
research grant to provide you with a stipend). It is a very good idea
to make sure there is more than one professor in the department
whose work you find interesting, in case things do not work out in
the lab of the first individual.

A second (and the most common) way to choose a school is by
the *reputation of the department.* A student should apply to the
best departments that he or she has a chance of getting into (and
one or two "safe schools"). There is a definite hierarchy of graduate departments, and the standing of your alma mater within this
hierarchy can determine the course of your career. The university
from which you received your Ph.D. will follow you everywhere;
it will be listed in university catalogues, mentioned when you
give a talk, asked for on grant applications, and inquired about
during casual conversations with colleagues. Graduating from a

prestigious department will help you to get jobs and rise in the profession. Furthermore, if you attend a good department, you will have a choice of many excellent professors to work for, and there will be sufficient money for research and financial aid. Signs of a strong department include the presence of many postdoctoral fellows (who are attracted to the department for further training), and the existence of government-bestowed training grants. The reputation of a department is synonymous with the reputation of the faculty members that comprise it. Superb researchers, who are making significant contributions to their field, make a superb department. Since the quality of the various departments within a university varies, the department, and not the overall university, is the critical unit of evaluation for graduate school. Unlike undergraduate students, who take advantage of the educational offerings of a number of departments, and are more concerned about the teaching abilities of their professors than their research reputations, graduate students generally deal solely with their own departments, and are mainly concerned about their professors' research standings. Qualities that determine the reputation of a research professor are discussed in detail in Chapter 2.

The Admissions Process

Undergraduates are in school to benefit themselves. Grad students, too, go to school for their own benefit, but many are accepted because they also have the potential to benefit the faculty. Faculty *need* graduate students. They need them to carry out their ideas, to perform their experiments. Grad students are the workhorses of academic departments. Admission to graduate school is thus a seller's market, with the student selling his/her talents, his/her ability to perform a professor's work. Thus, if you have plenty of *research experience* and a *good academic record,* you will get into a graduate school.

Unless a department or university specifically states that applications will not be reviewed until after a given date, you should assume that it has a rolling admissions policy. That is, an admissions committee reviews and acts upon applications as they are received (typically from November through February). Since the deadline for submitting your application is *not* synonymous with the time when the review process begins, it is to your distinct advantage to have all your application materials, including the letters of recommendation, sent in several weeks (or more) before this deadline.

Based on the admissions committee's review (the evaluation criteria are discussed below), applicants are ranked as: outstanding (invite for interview and admit if they "interview well"; qualified for admission (hold for future consideration pending the decision of the "outstanding candidates" to accept or reject their offer of admission); and poor (reject). Thus, it is only those applicants that look the best, at least on paper, who are invited (usually at the university's expense) to visit the university during February and/or early March preceding the start of the fall semester. Shortly after these visits, offers of admission are extended. You should know that if you do not receive such an invitation, you can always request a visit to the department at your own expense. As a result of this visit, you might be moved from the "qualified" to the "outstanding" category.

According to a current agreement among graduate institutions, applicants have until April 15 to accept or reject their offer of admission. You should be aware that departments of comparable quality in different universities may be in competition for some of the same students. Since each department has a limited number of openings available, and since the applicants that look great to one department will most probably also look great to another, a department must make more offers than it has slots available in order to fill its class. (The ratio of the number of offers to available slots depends on the quality and reputation of the department.) Once an admissions committee has extended all its available offers, it then has to wait patiently until it learns whether applicants accept or refuse the overture. Only after the admissions committee receives a refusal will it extend an offer to one or two of the next most qualified individuals on its list. Thus, as you are deliberating which offer to accept, be aware that if you wait until the April 15 deadline to refuse a department's offer, that department may not be able to fill its class with qualified candidates because those applicants in their "hold" category may have already accepted an admissions offer from another department. So be courteous both to your fellow applicants and to the departments that are waiting for your answer, and let those departments know your decision as soon as you have made it. On the other hand, if you choose to delay your final decision until the April 15 deadline based on the unlikely chance that you will receive an eleventh-hour acceptance from your first choice, it is your prerogative to do so.

Criteria for Admission

Admission committees use a number of measures to gauge the relative merits of applicants. The importance placed on each of the following criteria varies with the department. Some departments allow distinction in one measure to compensate for mediocrity in others. Some departments use cut-off scores on standardized tests and Grade Point Averages. *All* departments are impressed by research experience and knowledge of research techniques.

Letters of recommendation What other scientists think of your abilities and potential is of great interest to admissions committees. If you have performed research, your research advisor is an obvious person from whom you should request a letter of recommendation; in fact, the absence of such a letter will look suspicious. Your undergraduate advisor is also in a good position to judge your capabilities. A letter from a professor who knows you only as a grade in his/her course is less meaningful. If you are not sure if a potential reference knows you well enough to recommend you, come right out and ask: "Do you feel comfortable writing me a strong letter of recommendation?" Professors who do not feel qualified to evaluate you are relieved when given such an option. Remember that a mediocre letter of recommendation, which is interpreted by the admissions committee as a polite dodging of true opinion, is worse than no letter at all.

The Graduate Record Exam (GRE) Results of the GRE, a standardized test analogous to the SAT (Scholastic Aptitude Test), are required by most graduate programs. The exam contains a verbal, mathematical, and analytical section; specialized GRE "subject tests" (for example, in chemistry) are also required by some departments. The importance of the GRE varies. Some departments recognize that the ability of the exam to predict performance in graduate school is questionable; others rely heavily upon this easy standard of comparison as a matter of convenience. If your performance on the exam is below par for the program you are applying for, take the test again.

Transcripts Your undergraduate grades, especially in science and math courses, are reviewed carefully by admissions committees. Information on the range of Grade Point Averages (and GRE scores) of past entering classes of a particular graduate department is often listed in application materials, or can be obtained by calling the graduate admissions secretary of the department. You can use this information to compare your grades to those of students previously accepted for admission, and judge where you stand. If you received low grades during your freshman and sophomore years, your poor performance may be excused if grades rose significantly during your junior and senior years. The academic reputation of the school one attended is not overlooked—it is well recognized that a B from Princeton may be equivalent to an A from another college or university.

Students who feel that they are not qualified for admission to a Ph.D. program should consider trying to gain admission to a master's program. It is not uncommon for a late bloomer to use a good showing in a master's program as a springboard for admission to the department's doctoral program. Other than your money (M.S. candidates pay their own way), little will be lost by this maneuver, as the courses taken for the master's will satisfy many of the course requirements for the Ph.D. Moreover, if the student stays in the same lab, the research performed for the master's can be applied to the studies required for the doctorate.

The Essay Students can obtain an application packet by requesting one from the department in which they are interested; a general university catalogue from the campus admissions office will probably be sent at the same time. The application form contains a number of sections, most of which can be filled in simply and directly. However, the essay section, one of the more important and deciding parts of the application, is quite a bit more subjective and thought-provoking. Many prospective students have no idea what to put down here, and find the essay the most difficult and dreaded part of the application process. What types of things should one include in the essay? (1) If you have research experience, discuss the projects you worked on, the techniques you used, and the significance of the results you obtained; explain how participating in research "turned you on" to science, and what you learned from the experience. (2) Mention why you perceive yourself to be a good candidate—but be subtle, and don't brag. (3) Clarify why that particular department is attractive to you, and how your interests and those of some of the faculty coincide; if there is one particular faculty member whose research you find particularly interesting, say so, as the professor may fight for you and influence the committee's decision (do not explicitly exclude other professors, however, as there may not be room for you in the lab you want). (4) Disclose (but do not dwell on) things that may have shaped you and sparked your interest in science; perhaps a relative with a neurological disease influenced your decision to aspire to a career in neuroscience research. (5) If you wish to, briefly mention some of your interests outside of science. (6) If extenuating circumstances resulted in your having a poor academic record, disclose them (this may come up again at the interview), but do not harp on the issue; if you were a late bloomer in college, point out how your grades improved in your junior and senior years. (7) If your "significant other" is applying to another graduate program at the university, mention this, too; if you are an excellent candidate, your department may try to influence the other department so that both of you will be accepted.

Make sure that your essay is well written; if possible, have a professor that you know check it over for organization and content before you put it in the mail.

The Interview Potential graduate students asked to come for an interview are being evaluated to see if their interests, skills, and personalities are well matched to those of the faculty. Only students whom the department is seriously considering accepting are invited for an interview. This fact in itself should serve to relieve some anxiety if you are a nervous interviewee: if you are asked to come to the campus, the department thinks you are pretty good. Another tension-relieving realization is that the interview is two-sided: you are evaluating the department as much as the department is evaluating you.

An interview is not an occasion for jeans and T-shirts. Men should wear sports jackets (suits are "too much") and women should dress in a correspondingly appropriate manner. You probably will meet individually with several professors in your area of interest, spending fifteen minutes to a half hour in each private office. If a meeting with a particular professor that you are interested in has not been scheduled, ask if it can be arranged.

Each professor will conduct the interview somewhat differently. Some will immediately start explaining their own research interests, and never get around to asking you anything about yourself. Others will spend the entire time telling you the virtues of the city, the university, and the department. Still others will ply you with questions about your interests, your research background, your grades, and your goals. These interviews may appear to be spontaneous discussions, but the interviewers, even if seemingly casual, are seeking important information about you, and you definitely can and should prepare in advance for this. Make sure that you thoroughly understand any research that you performed (you will be caught if you try to bluff your way through an inquiry); organize your thoughts, and be able to discuss the significance of your work, and not just the techniques that you used. Be prepared to discuss the type of department or program you are looking for, and the research interests that you have. Think about your goals. Read some of the publications of your interviewers, so that you can ask *intelligent* questions about their work (you do not want to say, "Tell me about your research"), and can show your interest.

The responses of interviewers to potential students are subjective and complicated. As Dr. Roger A. Gorski, Professor of Neurobiology at the UCLA School of Medicine, points out below, much depends on how the professor views his role as an advisor.

♦ A professor looks for a number of obvious traits in a potential graduate student: a solid academic record particularly, but not exclusively, in courses relevant to the proposed degree program, competitive GRE scores, plus a commitment and dedication to research. A meaningful undergraduate (or master's) research experience that resulted in research publications is very impressive. However, there are two aspects of an applicant to which there is no set response: independence and a firm idea of one's research interests. The potential response to both of these is determined by the professor's attitude towards his/her research. If the professor is one who routinely *assigns* graduate students to carry out components of his/her funded research program, independence might be viewed as a great threat, but since research strategies do evolve, independence may be viewed as a great strength. A clear idea of one's research interest can be a weakness if they do not match the interests or expertise of a potential mentor and a strength if they do. A student who persists in wanting to continue in his/her undergraduate research field may be viewed as really interested in the field *or* too narrow minded and unlikely to expand and diversify. In any interpersonal relationship—and the student-mentor one is a prime example—personal chemistry is important but difficult to predict in advance. My advice: be comfortable with yourself.

What inquiries should *you* make during your visit so that you can adequately judge the department and the program? There are a number of issues that you should be concerned about: (1) What are the teaching requirements? Teaching a few undergraduate courses as a teaching assistant (TA) is a good learning experience, and is well worth the time spent away from research. However, a department that insists that all its students teach an excessive amount (more than two years, or more than one course a term, without specifically being salaried as a TA) is exploitive, and may be using its students to avoid hiring more faculty or taxing the present faculty. (2) How many new students is the department planning to accept? It is good to know how much competition you have so that you have an idea of where you stand. (3) How many of the students that typically start the program end up finishing? Some departments have a reputation for taking in far more students (sometimes using them as TAs) than they graduate. (4) Are the faculty in the department actively doing research, and are you interested in their research? There have to be well-funded labs working on contemporary issues for you to work in; you do not want a department heavy with inactive, older scientists, since research needs new blood. (5) Are the faculty congenial, or do they engage in infighting? A department in which the various labs collaborate, and/or offer assistance to each other, is most productive and, needless to say, most pleasant to work in. (6) What is the stability of the faculty? That is, over the next couple of years, are professors you are considering working under likely to retire, go on sabbatical, or leave the university for a position elsewhere? (7) Are there places for new students in the labs in which you are interested? You do not want to arrive and find out there is no one of interest to work for. (8) Are you guaranteed financial support (tuition waivers and stipend) throughout your training? (9) Is departmental funding available for students to attend conferences? Professional meetings are important educational events, and provide crucial networking opportunities, so students who are unable to attend will find themselves at a real disadvantage. (10) Is there sufficient funding for fieldwork? If applicable, you need to know if the expenses of fieldwork and/or travel can be covered. (11) How well equipped is the department? Today's science requires sophisticated, state-of-the-art technology; without adequate instrumentation, a researcher will be handicapped and frustrated.

The answers to all these inquiries are very important, so even if the university does not hold formal interviews, make sure that you visit the campus and speak to the students and faculty. If the department is really interested in you, you should be allowed to visit at any time.

During your visit, you will probably be treated to lunch, or perhaps even to dinner (such hospitality is indicative of the department's interest in recruiting you). At some point during your stay, *make sure you talk to the grad students without the faculty present.* Ask them what they think of the department, the program, and the individual professors. Are they happy there? Are they optimistic about their careers? The perspective of students is critical and very revealing.

The Unsuccessful Medical School Applicant

Some of the students presently in graduate school originally applied to medical school and were rejected. They are hoping that taking some graduate courses, publishing some papers, obtaining a master's, or even going all the way for the Ph.D. will bolster their credentials, and make their second application to medical school more successful than their first. Many departments will not mind if your objective is to go to medical school and you think getting an M.S. will help. They may also be agreeable if your intent is to do medical research (if they believe you), and you hope to get a M.D. after receiving your Ph.D. What *will* upset admissions committees and departments is recognizing that they are being used— that you plan only to take some courses or get a publication, and then vanish the moment a letter of acceptance from medical school arrives.

Financial Aid

There are a number of types and sources of financial aid available to Ph.D. students in the sciences, and most of these are based on merit, not need. In contrast to students in the humanities, the vast majority of Ph.D. students in scientific fields receive full tuition waivers, a stipend (currently between $12,000 and $16,000), and health-care benefits.

Individual professors or departments usually initiate the financial aid process by recommending students for awards available from the university or national agencies. Governmental agencies, such as the National Institutes of Health (NIH) and the National Science Foundation (NSF), sponsor training grants and fellowships to support graduate students. The fellowships ("individual predoctoral fellowships") are especially prestigious, and difficult to obtain. Securing a spot on a competitive training grant is also considered a distinction; academic departments are offered a finite number of these awards to give to students of their choice (if you are equivocal about your career direction, be careful that you do not get locked into a funding situation that requires "payback" after graduation in the form of a year or two of research activity). Note that such training grants are restricted to U.S. citizens or permanent resident aliens (i.e., "green-card" holders). Many universities sponsor fellowships of their own, although there are usually very few of these awards available, and they go to outstanding students only. A very common form of financial aid comes in the form of the research assistantship (RA), a position funded by a professor's grant (i.e., by money from NIH, NSF, etc.), and filled by the professor's students; for these awards, student recipients (RAs) are expected to perform experiments targeted in, or related to, the grant paying their salaries. Importantly, *experiments worked on under a research assistantship can be used to comprise a student's dissertation studies.* Research assistantships usually include tuition waivers, and pay enough to cover general living expenses. These assistantships do not have a citizenship requirement.

Teaching assistantships are another common form of financial assistance. A number of slots for these positions are awarded to individual departments by the university. Teaching assistants (TAs), who usually receive tuition waivers and earn a salary comparable to that of an RA, work hard for their money. TAing is supposed to consume no more than twenty hours per week of a student's life, but there may be times when more than this is required (especially if the student has to prepare lectures). A TA is required to attend the class that the professor he/she is assisting teaches, help the professor with paperwork and grading, and hold office hours for undergraduates that need assistance. Beginning TAs assist in lab courses and/or hold review/discussion sections; advanced TAs supervise lab courses or give lectures. New teaching assistants rarely have a background for these responsibilities. They probably have never taught before, and they are likely to have only recently completed the course they are now assisting in teaching (remember, first-year grad students were undergraduates just a few months earlier); some are assigned to courses that they have never even taken. However, TAs learn on the job, and manage to improve their technique with each class they assist. They sometimes keep only one step ahead of their undergraduate students in regard to conquering the material being taught, but one step is often all that is needed.

> ◆   For the most part, teaching was an enjoyable and valuable experience. I was always surprised at how much *time* it took, though. I gained competence by doing it. (Ph.D., Psychology, biopsychology research emphasis, University of California, Berkeley)
>
> ◆   TAing was great. I was nervous in the beginning, but then I realized the students didn't know any more than I did (I hadn't taken the class before I TA'd it—it was a medical school class). I helped in a lab; I didn't give lectures. The lab was held for four hours twice a week, and then there were times when I had to help small groups on weekends. The students who TA courses where they do a lot of grading of tests are the ones who seem to complain the most. (Graduate student, Neuroscience, University of Rochester)

If a department does not offer a stipend, students must take the initiative to research and apply for financial aid on their own. Many private foundations, and some governmental agencies, confer grants and fellowships to worthy students that fall into specific categories. Monies for minorities, women, children of veterans, students researching particular topics, older students, etc., are obtainable by those who are deserving and seek them out. A graduate student advisor or university librarian can provide you with information on the numerous and varied grants and fellowships available. You may also wish to inquire about "federally guaranteed loans," which are obtainable even if a student is receiving another form of financial support. This particular type of financial assistance is less desirable, however, as loans must eventually be repaid.

Foreign Applicants

> ◆   The U.S. has the most advanced scientific technology and the best scientific training. Most importantly, there is more freedom here to debate scientific issues so that you do not need to worry too much that challenging authority or dogma will damage your career. In contrast, my country does not provide facilities or faculty as good, and grad schools there have too much book-reading and examinations. That is not the type of training process I wanted, so I went overseas to find the best training system. (Graduate student, citizen of Taiwan)

♦ Why did I decide to go to graduate school in the U.S.? Let's put it this way. There were twenty-four biology-major students in my year. Seventeen of them are now in the U.S. doing their Ph.D.s. It is the main trend. Why is that? First, you can get better training here. Second, you can get a job more easily when you go back to your country. (Graduate student, citizen of Korea)

♦ I came here because the best schools in the world are in the U.S.A. The universities are very competitive and provide a lot of teaching aids. More specifically, grad schools here are very research-oriented. They can support students financially. They have all kinds of computing facilities which are fundamental for today's research. (Graduate student, citizen of Turkey)

If you live in a country outside the U.S. and are contemplating coming here for graduate study, do not worry about standing out or being "different"—many, many international students are walking around on American campuses. In fact, in many departments and fields, one-half of the graduate students are foreign. There are numerous sources of information, accessible from your country, on American universities and their application procedures. An initial step might be to speak to professors at your university who may be able to offer some advice. Libraries and bookstores will probably also have relevant information in the form of books and catalogues, and government education agencies will have pamphlets, books, and other materials. An office of the USIS (United States Information Service) can be found in all major cities of the world, and here you can pick up the free publication, *If You Want to Study in the United States;* books and printed materials on the various American universities and their costs are also available. *The International Student's Guide to the American University* by Gregory A. Barnes (see References) lists educational advising centers around the world, where information on American universities and application procedures can be obtained. This book also provides useful information on American cities and their climates, choosing schools and applying, student housing, visas, and American culture. Finally, if you can get on the internet, you can reach a homepage (http://www.chem.niihama-nct.ac.jp/univ-full.html) from which you can access information on universities throughout the world. Descriptions of campuses, departments and their requirements, professors and their research, course offerings, facilities, university cities, etc., are easily obtainable.

To be admitted to a graduate program in the U.S., you must have good grades and letters of recommendation, acceptable scores on the GRE exam, and scores on the TOEFL (Test of English as a Foreign Language) exam that reveal competency in the writing and reading of English. Since an important source of financial aid for foreign students at many institutions is the teaching assistantship, it is also important that you be proficient at *spoken* English, or you will find communicating with your students a struggle. If you do not attain this proficiency before you arrive on campus, ESL (English as a Second Language) teachers at the university will help you perfect your language and teaching abilities. Some universities may require you to take the Test of Spoken English (TSE) before granting you a teaching assistantship.

You should start investigating schools and arranging for application requirements over a year in advance—there are many things to be done. You can make things a bit easier by taking advantage of GRE Locator Services, a service that matches students with schools seeking applicants with their particular qualifications (consult an educational advising center about this). Also, be sure to check with the U.S. embassy in your country so that you can obtain the appropriate type of visa.

Foreign students may be amazed to find that the American university is like a small city, and students may not have to leave campus to find housing, food, entertainment, clothes, school supplies, or perhaps even a bank or post office. Despite these conveniences, international students will initially still find themselves confused and stressed by the cultural differences they encounter. This is discussed in Chapter 13. The foreign student advisor on campus should be able to help with some of the difficulties experienced, while support from others sharing this cultural shock can be found at the foreign student association on campus.

Foreign students who have difficulty getting accepted into the Ph.D. program of a choice department should consider coming to the U.S. for a master's degree. After they obtain this degree and have "proven themselves," they can then apply to a better department for the Ph.D. Foreigners who hold an M.D. degree may choose to skip the Ph.D. route altogether, and try for a postdoctoral position in the U.S.; an M.D. with some productive years of postdoctoral experience can readily compete with Ph.D.s for research positions.

It has become a common ruse for foreign students unqualified for the better Ph.D.-awarding departments to enroll in lesser-known ones for the sole purpose of enhancing their mastery of English and bolstering their knowledge—all the while receiving financial support. While still registered in these departments, these students then apply to better programs without informing their departments that they are doing so. Students taking advantage of departments in this way should beware: American institutions are cognizant that these maneuvers are being attempted, and are screening their applicants carefully. Students intent on enrolling in a Ph.D. program at a new department after completing a master's or a stage of the Ph.D. should be sure to obtain a letter of recommendation from their lab head; an application without such a letter will seem most suspicious, and may prompt the applicant's rejection. Remember that all it takes to uncover this ruse is a telephone call from a member of the admissions committee to a colleague in the department from which you wish to transfer.

2

Selecting an Advisor:
Whose Lab Is Right for Me?

As a new graduate student arriving at the university, you are likely to be somewhat bewildered as to what to expect, and what is expected of you. As the yearly incoming class in a graduate department is often small (do not be surprised if you are one of a handful), you are unlikely to feel any class camaraderie behind you. To ease some of your perplexity, orientation meetings may be scheduled to welcome you and help you get started, or/and you will be sent to speak to an individual, either a faculty member, or a full-time assistant of grad students, who will inform you of course requirements, fellowship and grant opportunities, and other essential information.

You may also be informed that you are expected to start participating in research right away; many departments do indeed introduce their students to the world of research at the start of the first year. Other departments, however, wait until students have finished their classes, usually during the second or third year, before expecting them to engage in research. Either way, there are two means by which students are assimilated into the research environment. Some departments expect the student to take the initiative—after talking to various professors and reading up on their research, the student delineates his/her own interests, decides which lab will be best for him or her, and asks permission of the lab head to join that lab to pursue his or her doctoral work.

♦ At Brown, we could have started doing research any time after we entered grad school—if we found a professor who would take us on. Since our qualifying exams were very important, many profs said to forget about research until after they were over. When mine were over (in my second year), I knew I had to find someone to work with. I approached a professor who did theoretical stuff. He wasn't sure if he wanted to take me on, but he said I could work for him for that summer. I did, and ended up staying—and publishing four papers. Some other professors refused me because they had no money. (Ph.D., Theoretical Physics, Brown University)

Another popular means of assimilation in laboratory science departments is the lab rotation system. Here, students spend several weeks to as much as a semester in a laboratory of their choice, where they watch others and carry out small projects; in succeeding periods, the students move on to other laboratories, each with a different research focus. After completing a number of such rotations, students typically form a pretty good idea of what they are interested in, and they may have had enough contact with lab heads to decide who would be a good supervisor for them (lab heads evaluate the *students* during this period, too). They are then in a relatively good position to make the crucial decision of with whom to work for their doctoral research. This is a matter not to be taken lightly, however, and students have to do further homework.

♦ Except in unusual circumstances, it is a plus to choose a program in which the students do rotations through a few labs before choosing the one they want to stick with. While this does decrease the total amount of good research an individual is able to carry out during graduate school, the benefits outweigh the costs. Many advisors who vehemently disagree with such a program only worry that students will not want to stay in their labs. It is sometimes difficult making a graceful exit from a laboratory one might want to leave if things are not working out. A program which includes rotations may help a new student avoid becoming ensnared in a very difficult situation. (Ph.D., university/department withheld upon request)

♦ Don't be afraid to experience new things. Most people entering graduate school have firm ideas about what research areas they are interested in. For two of my lab rotations, I did experiments in a reproductive endocrinology lab and a developmental biology lab even though my main area of interest is the study of human disease. I had great experiences and learned a lot of new information, and this is important since this is your last chance to do something completely different before beginning your specialized thesis research. You may be surprised about what you can become interested in! (Graduate student, Biology, University of Virginia)

Which Advisor Is Right for Me?

Picking an advisor, the professor/mentor whose lab the student will spend the next four or more years working in, is surely the most important and consequential decision that a graduate student is forced to make. This crucial decision will largely determine how enjoyable or horrific graduate training will be, and will profoundly influence the chances of eventually pursuing a career in science. Unfortunately, it is a decision that many new students are *ill-prepared to make.*

♦ In looking back at the way in which I chose where to go for graduate school and which lab to work in, I absolutely cringe at my extreme naiveté. I carried the mistaken assumption that most graduate programs in my field would be roughly comparable in terms of how hard I would work, how long it would take me to get out, and how I would be treated by my advisor. I guess that I had pictured the process of getting a Ph.D. as being similar to the far more regimented process of getting a medical degree. Thus, I was astounded when I started graduate school and noted the extreme variability, in the things mentioned, between Ph.D. programs in different departments, and *even between different labs in the same department.* (Ph.D., university withheld upon request)

One comes to graduate school eager, curious, and idealistic, with high hopes and ambitions. A beginning student, however, is likely to be unaware of the factors that will either ease the way to professional success or pile obstacles in the path; he or she is ignorant of the determinants that will eventually make or break a career. Many of these factors are linked inexorably to the person that the student chooses to work for. A good advisor can lead you gently down the road to achievement and self-realization during your graduate years, and then later, with his or her influence, can open new paths for you as you progress to the job market; a poor advisor can halt your professional journey almost as soon as it begins. Thus the picking of an advisor is a crucial decision whose importance in determining future success or failure cannot be overemphasized. It is a decision to be agonized over.

♦   I remember attending an orientation party for new students. I overheard some of the advanced students talking about one of the new students who had decided to work in the lab of one particular professor. Obviously this prof was not a good choice, as the students were actually taking *bets* as to how long the student was going to last in the lab! (Ph.D., department withheld upon request, University of California, Los Angeles)

Factors to Consider

Research Interests The initial paring down of potential mentors can be made simply on the basis of mutual interests. Can you imagine spending four or more years of your life doing research on a subject that is of no or little interest to you? The doing of science requires so much time, attention, creativity, effort and contemplation, that without the necessary love of subject matter to drive you on, the work will not get done. Thus, in addition to the emotional benefits derived from doing something that you like, it is also of practical necessity to pick a lab whose research focus matches your interests. Talk to students about their research. Go to the library or go on-line, and read the recent publications of advisors that you are considering working for to see if you find their work exciting. (Resources such as Medline [http://www.ncbi.nlm.nih.gov/PubMed/] are available on the Web, allowing you to look up abstracts of a professor's papers by doing a search of his/her name on your computer.) Make appointments to speak to the professors about their work and to ask relevant questions; they should be pleased that you have shown an interest.

If the research there is to your liking, *it is a smart move to go into a lab that is working on "hot" topics and/or using "hot" techniques.* Fashionable topics and methods are so because they hold the promise of revealing important truths. However, views and technologies change, and what is hot today, may be lukewarm or cold tomorrow (and vice versa)—it is very hard to tell. Still, speak to scientists in your department to see what they think the well-funded areas and useful techniques of the near future will be. What research will have applications in industry if you decide to leave academia? What research has direct clinical applicability, and thus may have an edge for future funding? Is the research in the lab you are considering significant, or is its meaning questionable? These may seem like pragmatic points to a new student fascinated by a particular research topic, but the issues are critical. Read Chapter 16 before you make a decision.

Compatibility No one wants to work for someone they do not like. For graduate students, who will be dependent upon their lab head for scientific advice, counsel, direction, encouragement, and letters of recommendation, this is an especially important issue. Furthermore, advisors can often protect their students from certain departmental decisions or rules, and secure for them financial assistance in the form of fellowships, a slot on a training grant, research assistantships, or teaching assistantships. If the situation arises, advisors can also help promote their students by inviting them to coauthor book chapters or journal review articles. Thus, you want an advisor who genuinely cares about you and your career. Selecting an advisor whose personality is likely to clash with yours is not a wise idea!

Being "buddies" with your advisor may make life easier and the working environment warm and comfortable, but this degree of friendship is not really necessary; mutual respect is a more important requisite and, combined with a pleasant working relationship, should suffice for most students. Everyone is different, however, and what suits one student may be distressing for another.

◆   One's relationship with an advisor is a very individual thing, and it's hard to generalize about how to choose one. I narrowed the candidates based on my affinity for their research area, but made my final choice based mostly on personality. I finally chose a rising young assistant professor over two well-established mid-career full professors because he seemed so much more excited about his work. I figured either he would be great or he would drive me crazy, and it was worth the chance; as it turned out, he only occasionally drove me crazy, and his energy and enthusiasm were great for me. I worked hard and learned quickly, but at that age I tended to be overly sensitive and to lack confidence, and the most important thing I got from my advisor, besides of course actual scientific advice, was lots of encouragement. In my case, my advisor was an advisor in the true sense of the word. I got plenty of help from other students in the group, and had very good scientific interactions which included a couple of collaborative papers with another professor in the department, but it was clear that my advisor was providing most of my guidance. (Ph.D., Chemistry, University of California, Berkeley)

With the proper advisor, the laboratory can actually be a most enjoyable place. This professor states his philosophy of management:

♦ I believe that people should have fun in the lab. There should be that proper balance of work and fun, total concentration interspersed with lighter moments. A professor and student should have a match of personalities, a similar philosophy of life. (Ph.D., Biophysics, University of Rochester School of Medicine)

Some students can tolerate an advisor's "disagreeable personality" as long as the research is promising and exciting; others cannot. Prospective lab members should thus spend time talking and interacting with professors under consideration so that potential personality problems can be foreseen. Do not forget that all types of people become professors, and that even though a person may deserve respect based on research, it does not necessarily mean that he or she is someone that you want to spend time with.

♦ I might not have known how to pick an advisor but I knew what project I wanted to work on and that governed my decision. I figured I better enjoy the project because I was going to spend 4–5 years working on it! Looking back I would have also thought about my advisor in that sense too. I always tell people, "You had better like your advisor because you are going to have to work with him/her for the next several years and then you will need letters of recommendation, etc." I think that the advisor-grad student relationship governs your grad school experience. Most people that like their advisor like (or do not hate) grad school, but all of the people that I know that didn't like their advisor hated grad school. I got along great with my advisor and I had a lot of fun in grad school. (Ph.D., Chemistry, University of Rochester)

Even when students and professors seem to "click," however, final decisions should never be made without further consultation. Never, never agree to enter a lab without first speaking to upper-level graduate students or postdocs in the department; if possible, speak to the present and former students of the professor in question. *These people, who are often willing to speak freely and openly, are your most important source of nitty-gritty information about faculty members.* It is imperative that you make use of this extraordinary resource.

Questions to Ask

The following are topics that you should bring up with graduate students or postdocs.

Mentor's scientific reputation All things considered, it will be of enormous advantage to work for the best and most active person in your area of interest that you can. A prolific and superior scientist is unlikely to allow poor research to be performed in her laboratory, and thus, as this person's student, you are apt to be trained carefully and appropriately. Importantly, this observation is not lost on future employers when you apply for a postdoctoral or faculty position. As part of the application process, you will always be asked in whose lab you were trained, and your mentor's reputation, good or bad, in a way becomes your own. A letter of recommendation from an advisor who is respected in the field carries much weight, whereas one from an advisor whose work is considered below par may carry none at all. Furthermore, a well-known scientist is usually also well connected; he is in a position to recommend you for postgraduate employment positions with friends who are also renowned in the field. Thus, an advisor in excellent standing in the academic community can open up many doors for you, whereas one with a poor reputation can actually hinder your chances of becoming a career scientist, no matter how talented you may be.

On the down side, the more "famous" professors are also the busiest, giving invited talks out of town, serving on study sections, chairing committees, etc. You may have trouble getting their time and attention. They also may feel so comfortable with their established reputations, that they cease to be hungry to publish, creating for their students a deadly environment lacking in focus. Furthermore, there is no guarantee that they will be good "mentors," in the true sense of the word.

It may surprise some of you to learn that even the best research institutions contain professors who are doing poor or yesterday's science. Typically, these are tenured faculty who did some good work in their earlier years, but have been unable to keep up with the more recent advances and technologies. Others may have snuck into academia years ago when positions were easier to get and somehow have managed to hang on, perhaps by publishing in mediocre journals. Such scientists may have little or no "pull" or power, and may not be in a position to help you with your career. A student needs to be in an exciting and productive lab, so make a point of noting whether the advisors you are considering publish often in respected journals, if they have published in the last year or so, and if their research is well received by others in the field. Beware of those that have produced nothing but review articles for the last few years; this may indicate that their labs have stopped being productive. Also watch out for those whose names consistently appear in the middle of the list of authors, a sign that the projects did not stem from their labs, and that they played a relatively subsidiary role in the work.

> ♦   The faculty members in my department ran the gamut; some were good and some weren't, but I do not think I was able to judge them at the time. Looking back on it now, I see that some people were doing a great job and there were others who were just getting by. (Ph.D., Biochemistry, State University of New York)

The scientific reputations of faculty members are well known to advanced students and postdocs, so speak to them. *They can tell you who are the respected researchers, and who are considered dead wood.*

The style of the professor and his/her lab Visit the laboratories in which you are considering working. Every lab has its own style, and one style may appeal to you more than another. In some labs, researchers work closely together in an environment of enthusiasm, friendship, and congeniality; the members sometimes chat while they perform routine manipulations, and humor adds levity to the atmosphere. Other labs are more solemn, and may appeal to those who prefer to work in a quieter environment. The pervasive aura is very often influenced by the personality of the lab head, even if his or her actual presence in the lab is minimal.

The personality of the professor also affects many other aspects of lab life (or field life, for those doing fieldwork). Some advisors are very directive and controlling; they may demand long hours in the lab, or they may give students little leeway in picking their own projects. Other advisors are more flexible; they allow and may even encourage students to design their own research, prefer to let students figure things out on their own, and allow them to use their time as they see fit.

♦   My graduate school experience and my postdoc experience
were opposite extremes. In grad school, I decided how, when, and
how often I did experiments, and how the data were analyzed. My
advisor seldom asked me what was going on; maybe once a month
he'd say "So, how's it going?" Looking back, although I did like
him, it would have been nice to have a little bit more involvement. I kept *him* informed about what I was doing. He never
pushed, and sometimes I wished that he had a bit; I think I needed
that kind of guidance. He would come into my office and chat on
a daily basis, but it wasn't about science. Also, we didn't have lab
meetings. My postdoc advisor was very different. He told me
every single day, probably every half hour, what I needed to do,
and he would check on me often [Note: this is very unusual]. He
wanted to be in total control. He told me how to hold a slide, how
to fold computer paper. This was not good—if anything, I think I
would have preferred the two situations to be reversed, total
structure first and then the freedom. On the other hand, if I had
the excessive structure first, I might not have finished my Ph.D.—
it was way too overbearing. I do not think most advisors are like
these two—most are probably somewhere in the middle. (Ph.D.,
Biology, University of Texas)

Some advisors are unusually critical and demanding, and their
labs may be somewhat tension filled, but they may force students
to do their best and be most productive. Other advisors may guide
in a gentler fashion, being known for their understanding, support, and helpfulness. Some advisors actively participate in lab
work, and are a presence in the lab; others have given up hands-on
research entirely. The best advisor is also a good supervisor—
someone who will guide, motivate, and encourage, and who will
direct the lab in a productive and efficient way.

♦   My advisor said that if I wanted to be successful, I had to work
hard and outshine others. He was very demanding—students in
other labs called him a slave driver. Initially, we felt that our lab
was our jail. But when we completed graduate studies, we were
glad that he drove us very hard: I do not think he could have made
it pleasant and gotten us to be as productive as we were. At the
same time, we had plenty of fun outside the laboratory. I consider
myself lucky to have come out of that lab—I would advise everyone to go through that kind of system. My advisor prepared me for
what I was going to expect after graduate student life; he made me
analyze and think more than I was used to doing. He taught me
how to give a seminar, review articles, write grants, and conduct

experiments. He stressed the importance of avoiding variability in doing experiments and in staying focused on our projects. He was actively involved in the conducting of experiments and made us critically analyze the results. Some of the students did not like this way of life and one among them left the lab. (Ph.D., Physiology, Kansas State University School of Veterinary Sciences)

Perhaps the most important personality trait that one should inquire about and seek out in an advisor is fairness. A fair advisor is aware of her responsibilities to a student that she takes on, does not make unrealistic demands of her student's time or efforts, does not change her mind midstream, does not make use of her student to the student's own detriment, and is honest with her student, evoking trust.

Funding How well funded your prospective advisor is is a critically important issue. Money is always the bottom line in science, and even well-known researchers can have their work come to a halt if a grant does not get funded. Most professors will not take you on if they know that they cannot support you or your research. It is probably not advisable to enter a lab where the funding situation is tenuous, for example, where the professor's one and only grant is in its last year of funding, but he "hopes" to get it renewed. If a professor has, say, two years of grant support left, check to see if the department has budgeted itself so that it can pay for your research after this time if no new grant money comes in.

♦   Ask the potential advisor about grants. Does the lab have current funding? How about in the recent past? Are grant applications currently submitted or in preparation? A history of poor funding for a laboratory bodes ill for anyone considering coming on board, whether the University will pay the student's stipend or not. (Ph.D., university/department withheld upon request)

A poorly funded laboratory—say, one that is relying on a single, small grant—is not likely to have the latest equipment and may be low on supplies, making ambitious research projects difficult, if not impossible, to carry out. Such a lab is unlikely to have technicians around to help with the work, and thus there may be more mundane chores than usual that you will have to do yourself (although you will certainly learn from these extra duties). There may be no funds for travel to distant field sites or professional meetings. Your choice of a research topic in this lab may be severely restricted to that covered by the grant, and you are not likely to have much freedom to explore new or different research areas.

Fate of graduate students The fate of former graduate students can be used as a telltale sign of which laboratory to join or pass up. If these students landed good postdoctoral positions in the labs of well-known researchers, or are now professors at major universities, then the system is alive and working well, and you can bet that their former advisors are capable and respected, and proven trainers of good scientists. These advisors are probably willing to give their students credit for their work and thoughts, and knowing how important it is for students to publish, are likely unhesitant about awarding their disciples first authorship on papers for which they did most of the work. These professors are surely also doing the job of networking for their people: they have written letters and made calls praising their students' virtues, and have used their influence in other ways to help their students succeed. These professors know their roles as advisors: they will train you well, and they will help you to get a good postdoctoral position.

For students with advisors like these, the need for a book of this sort is lessened: they are sure to be enlightened about the "rules of the game."

♦   As I look back, I can see that certain professors have seeded a whole line of students that end up in prestigious faculty positions, while other professors have few students that even remain in the field. Part of this must be attributed to the students themselves—possibly some professors tend to attract good students. I am sure, however, that the professors play a sizable part—some just use their contacts more and see it as their responsibility to help their students succeed. Placing students successfully makes professors look good, and increases the size of their dynasty. (Ph.D., Physiology, University of California, Los Angeles)

♦   Be sure to find an advisor who will later go to bat for you and who shows general support for ex-students in what is an increasingly difficult job market. Be wary of those who are seemingly unaware of the current difficulties in finding employment opportunities in academia, or who deny that there is any such problem. Be wary of those who seem to feel that when you are finished working in their labs, that it has been your "good fortune" to work for them, and that you should expect little else from them. (Ph.D., university/department withheld upon request)

Length of time to the degree Some students have slow starts. Before they find viable dissertation projects, they can spend years working on projects that never really pan out. As a result, it takes them a long time to finally get their degrees. These occurrences are not uncommon, and since exactly which experiments will or will not fly often cannot be foretold, the advising professors are usually not at fault. Sometimes, however, professors *are* to blame for students' lengthy stays in graduate school. They might have given the students misguided advice as to which projects were realistic enough to pursue for a dissertation. They may have been so aloof that the students floundered helplessly around on their own for long periods. Or, they might have encouraged the students to stay with unyielding projects for unreasonable amounts of time before finally admitting that the projects should be given up. Even worse, the professors may have been reluctant to let those valuable extra pairs of hands finish up and leave, insisting that the students do experiment after experiment, even though a sufficient amount of work for the dissertation had already been completed.

How long it takes the typical student to finish in a laboratory that you are considering entering is certainly something that you should inquire about.

♦ I feel that I was lucky in terms of how things ended up for me. I could very easily have been one of the many I saw who ended up being stuck in a lab where either (a) the advisor kept his students around for excessive lengths of time, or (b) the advisor required seemingly little of the students during an unusually short period of time, and sent his or her Ph.D.s out into the workforce ill-prepared to land a job in either academia or industry. In either of the above situations, it appeared to me as though the advisor cared little for the well-being of his or her students, but, rather, used them as mere tools to be cast away when they were no longer of use. (Ph.D., university/department withheld upon request)

♦ I guess the best advice I could give a student starting a Ph.D. program would be to find a mentor that you can get along with and most importantly, find a project that you can clearly see has a beginning, middle, and end. One way to do this is to interview mentors to see how long other graduate students have taken to finish, or ask other students. There are always mentors that have reputations of "never letting their grad students go." (Ph.D., Microbiology, University of Pennsylvania)

Lab size Professors who run large labs, with many students and postdocs, are usually doing something right. They surely are well funded, and are obviously researching attractive topics. Their labs are doubtlessly well equipped, and may have a solid technical staff. Students in such labs hold the advantage of having a large number of persons from which they can learn, people with a variety of backgrounds, skills, and ideas. Yet, there are also disadvantages to large labs. Students are likely to have to compete for their professor's time and attention, and may feel lost in the crowd. They will probably have little space of their own. They will have to share equipment with all the lab members, and will be able to use certain pieces only when they are free.

There are also advantages and disadvantages to working for an advisor who runs a small lab. Small labs have a more intimate atmosphere, where students tend to have easier access to their advisor and to other lab members. Individuals working in a small lab will likely have more working space, and fewer problems with the sharing of equipment and supplies, than their counterparts in a larger lab. They will also be less exposed to hustle and bustle, and can work in relative quiet. There is no place to "hide" in a small lab, however, and one tends to stand out there. There are also fewer people to discuss science with and, thus, fewer sources of new ideas, information, and aid. The variety of laboratory equipment may be minimal, although this is not necessarily so.

A laboratory with few members may be indicative of poor funding, an unfashionable area of research, a bland or narrow research focus, or the lab head's disagreeable personality. On the other hand, smallness certainly does not always connote negativity. Many professors prefer to keep their labs small so that they can more closely supervise the students, and keep the atmosphere intimate. Sometimes a lab is small because of limited, albeit adequate, funding, with sufficient money around to support a moderate number of students; this condition may not necessarily preclude academic excellence in these times of tight budget restrictions. A lab may also be small because a number of students have recently gotten their degrees and left, or because the professor is new to the university and has not yet gotten the lab totally up and running. *It is only by speaking to other grad students and postdocs in the department that you will be able to ascertain the information needed to sufficiently judge a particular lab under consideration.*

Labs that are totally unpopulated by grad students or postdocs may be so simply because of an unusual or highly specific research focus that is not attractive to most people. Students who like to work alone may be happy there. Still, it is probably wise to mark such laboratories with red flags until you inquire about the situation. The condition could indicate that the lab heads have severe personality problems, or are scientifically incompetent, and the word may have "gotten around." One of the authors remembers a professor who had a world-renowned reputation and numerous grants, but only one student passed through her lab over a span of six years, and that one lasted only six months. The woman was impossible to work for, although you would never know it by meeting her only once or twice. Hired technicians did all her research.

Tenure status: Working for new professors You may find that one of the new, assistant professors in your department is doing the research that strikes you as most interesting. This is understandable, as young profs are often hired because they are working in some hot, new area that is likely to yield promising results and exciting advances. It is tempting to consider working for one of these energetic hot shots and, indeed, getting your training in such a lab does have distinct advantages. New professors are highly motivated, up on the latest techniques, closer to your age, and have the potential to become well known in their fields (and since you will be publishing with them, much of that glory will go to you, too); they also will probably be around more (perhaps even working next to you at the bench) than a seasoned investigator who has more responsibilities outside of the lab and university.

The main disadvantage of working with new or relatively new academicians is that they have yet to prove themselves. It could turn out that they do not live up to their potential, and they may have trouble getting grants; they would then be unable to support your continued research. They may also not receive tenure, and be forced to leave the university before you are finished with your dissertation research; you would then have to find someone else to work for. Because the pressure to publish weighs heavily on new professors, they may be tempted to work you too hard, to have unrealistic demands, or to pursue low-risk projects that are insignificant yet will get them quick publications; they also may be so concerned about their own future that they will devote little effort towards helping yours. This could foster an attitude of competition between the two of you, and you may be given little credit for your work or your ideas. Also, new profs likely have little experience managing laboratories and the people in them (these things are never taught in graduate school, and are not requirements for getting a faculty position), and they may have difficulties in this domain. Importantly, these professors are unknown personalities to students, and so you will be unable to garner much information about them via this route. Working with a new faculty member is thus risky, but if you are very interested in the research, it may just be worth the risk. Remember that all senior faculty members were once junior faculty.

Leaving One Lab for Another

Because of interest changes, personality conflicts, or other issues, a student who has spent a year or so in one lab may decide it is best to move to another. This move is likely more anxiety producing for the student than for the professor. The student should not view the months or years spent in the first lab as wasted time; lab experience was gained, new techniques were learned, and focus likely increased during the "ill-fated" stay. It is unlikely that the department will think negatively about the transition; professors remember their own periods of shifting interests or situational conflicts. A number of lab switches, however, indicate a trend, and may be viewed more severely.

3

The Stages of Graduate School

This chapter provides an overview of the graduate school process. Succeeding chapters will explore most of these topics again in depth.

Coursework

The first few years of graduate school are different in many ways from those later on. This is because required courses are taken during these beginning years, and considerable time will be spent attending classes, studying, doing problem sets, and writing papers. In some departments, research does not begin until classwork is over. In others, research and classwork are carried out concurrently, and students must manage their time carefully.

Qualifying Exams

Shortly after students have completed their required coursework (usually in their second year), they face their biggest challenge up to this point: the passing of the "qualifying" (also called "comprehensive" or "preliminary") examination. This exam encompasses material from required coursework, and sometimes other sources, as it is considered a test of the entire field; a student's specialty area may also be covered. Every department administers its own version of the qualifying exam, and they can differ considerably. Some are in the form of challenging essay questions that must be completed in a circumscribed number of days, and require contemplation and extensive library research. Some are essay exams that must be completed in one day while in the confines of the testing room. Some are part written and part oral. Some are part essay and part literature review of a selected topic. Some consist of questions taken from material on a reading list submitted by faculty members. Some, especially those in departments of the physical sciences, consist of problem sets administered over a multi-day period. All require the ability to remember, assimilate, synthesize, and integrate a vast amount of material. If students are conducting research at this time, many stop for months to prepare; others continue what they are doing and study at night or during other free hours.

♦   There was a lot of anxiety over the comprehensive exam, which was written and oral and covered all our coursework. The written part was all day long for 4–5 days. One day was gross anatomy, the next day was histology, the next was neuroanatomy, etc. For a part of the oral exam, one professor came in with a heart, and I had to pretend for him that I was teaching its anatomy and physiology to a student. I *went away for three months* just to study! I would wake up in the morning and not finish studying until 10–11 at night. When I came back, I found I was the only student who was taking the exam! The six others all dropped out— they didn't want to take the exam. They weren't going to make much money in science, so some went to dental school, vet school, etc. They just sorta peeled off that summer. (Ph.D., Anatomy, Tufts University School of Medicine)

♦   For my qualifying exam, we were given, beforehand, a long reading list of articles that focused on the exam committee's areas of research. Much of the information was not covered in classes. The exam consisted of three essay questions based on the articles, and took 3–5 hours. A second part of the exam consisted of doing an extensive literature review of a topic of our choice. This was done on our own time. There was nothing at all tricky about the exam, and it was not associated with great anxiety. I guess they figured that if we were good enough to get into this selective department, then the weeding out was already done. (Ph.D., Psychology, biopsychology research emphasis, University of California, Los Angeles)

To study for the comprehensive exam, students definitely should try to get copies of old exams (some departments keep them on file), or they should query advanced students about the questions that were asked in years past. The structure of the exam rarely changes radically from year to year, and it is often possible to anticipate questions, or at least blocks of material, that will be encountered; prepare answers to those questions that you suspect will be asked. Keep in mind that *the research areas of the faculty members on the exam committee are likely to be topics on the exam;* professors rarely ask questions on areas outside of their own expertise. (Some departments have done away with these types of qualifying exams altogether, replacing them with a written or oral defense of a thesis proposal—see the Preliminary Oral Exam below.)

Advancement to Candidacy

If a student successfully completes all coursework and passes the qualifying exam, he or she has made it through the "trial period," and is considered to have advanced to the next stage of the academic process, becoming what is known as a Ph.D. "Candidate." Such a person is "only" a dissertation away from receiving the doctorate.

Students who flunk their "quals"—and at some schools there are many who fall into this category—are often given a second chance to try to pass another version of the exam at a later date. Those who fail this second attempt are usually asked to leave the program. Some students, realizing that graduate school is not for them, leave the department before they have to struggle with the qualifying exam; those that take the exam and pass and then decide to leave are often awarded a master's degree.

♦ The quals are the initial cut, the major hurdle. My exam was very difficult. I didn't pass the first time and had to take it again. The first time I spent a lot of time studying books; the second time I was smarter and just worked problems. You get used to working problems if you study this way, and the exam was all problems. We had two days of five hours of exams, followed by a two-hour oral exam. We had to study electromagnetism, quantum mechanics, etc. They could ask anything they wanted. A professor said "You would never learn this much physics if we didn't force you to." I don't think there is any strong correlation between how people do on the exam and how people do later on, but I would have had to leave if I flunked the second time. (Ph.D., Physics, Brown University)

Some departments use the qualifying exam as a means of weeding out excess students. At large universities with numerous undergraduates to teach, the graduate students are indispensable in their roles as teaching assistants. In some departments of such schools, a large number of graduate students are accepted each fall, exploited as teachers for the first year or two, and then the bottom half of the class is weeded out by a qualifying exam that is graded on a curve. This is not a good situation to be in, and is one of the reasons students shopping around for a graduate department should inquire about the ratio of students accepted to students finishing the program.

Getting Preliminary Data

The student's goal at this point is to come up with a plan for a series of studies that will suffice as the dissertation research. (Some definitions are in order here. The research that results in the granting of a master's degree is called the thesis research, and it is eventually written up in a manuscript that is called the thesis; the research that results in the granting of a Ph.D. degree is called the dissertation research, the written version of which is called the dissertation. At some universities, the words thesis and dissertation are used interchangeably.) Coming up with ideas for dissertation-quality studies is not easy. The first task for the student is to get some "preliminary data" showing that he or she is "onto something" (i.e., a line of research has been delineated that is doable and that is already yielding, if not interesting, at least interpretable, results).

Writing the Dissertation Proposal

After the student has conferred with his or her advisor and they have agreed on a series of experiments that will comprise the dissertation, the student must compose a formal dissertation proposal. In it, the student must relate: (a) the background of the proposed research in the form of a literature review; (b) the reasons why the work is important and what it will add to the field; and (c) a detailed description of the design of the experiments and the methods to be used. In fact, the dissertation proposal is similar in many ways to a small grant proposal. Some departments do not require a written proposal; in that case, the student merely presents his or her ideas orally at the preliminary oral exam.

Defending the Proposal: The Preliminary Oral Exam

A copy of the proposal is given to each member of the student's "dissertation committee," a group of four or five professors, usually picked by the student, that will make the final judgment as to whether or not the student passes the exam (this committee usually, but not necessarily, is the same one that judges the final oral exam). At the "preliminary oral exam" (the proposal defense), the student formally presents the proposed research to his or her committee, and the committee, for its part, drills the student with pertinent questions designed to determine the student's knowledge of the work about to be undertaken, and the soundness of the experiments involved. The committee may make changes to the student's original research plan at this time or, if the student or the plan are sorely deficient, may actually flunk the student. Students who fail the preliminary oral are often given a second chance at a later date; if their performance and proposal are abysmal, however, they may be asked to leave the program.

The Pure Research Years

If the student passes the preliminary oral exam, he or she then spends a number of years doing very little else but the dissertation research; these years are the heart of graduate school. In some departments, the committee will meet with the student at least once per year to evaluate progress, make suggestions, and solve problems. If the student is fortunate, some significant findings will be made during these years that are publishable, and he or she will thus be at least partly engaged in the challenging task of writing journal articles.

Writing the Dissertation

After the student has performed all the agreed-upon experiments/ studies, he or she is faced with the enormous task of writing about these investigations in a specified format. Together with an introduction consisting largely of a literature review, and an ending chapter of general discussion/conclusions, the tome that is created (often hundreds of pages in length) is the version of the dissertation that is presented to each member of the student's committee.

Defending the Dissertation: The Final Oral Exam

About two weeks after the dissertation is distributed, and after the committee has reviewed it, the student once again appears formally before the committee, this time for the "final oral exam," also known as "the defense." The committee questions the student on his or her dissertation, and decides whether the student will now be granted the Ph.D. The defense is often immediately preceded by a public lecture by the student on his or her research. The student is unofficially considered a Ph.D. after successfully completing the final oral.

Making Final Changes in the Dissertation and Submitting It

Committee members often insist that some minor changes be made in the writing of the dissertation, or in the analyses of the data. After tying up these loose ends, and making sure that the dissertation format meets official specifications, the student hands in the final version to the university and is officially finished with the program.

A general characteristic of graduate school is that *each student proceeds at his own rate,* a pace that is determined by the nature of the work and the personalities of the student and the advisor. Thus, every student reaches each stage of the process on a different time schedule. Being a year or two behind another student who started with you usually means little or nothing.

4

Classes, Journal Clubs, Lab Meetings, and Seminars

While a first- or second-year graduate student may spend a lot of time working on research, he or she is also expected to partake in a number of important activities away from the field or lab bench. Of these, attending classes will be the most time-consuming, although the other activities, including seminars, lab meetings, and journal clubs, which continue throughout the graduate years, are equally important for his education. It is only through many fronts that a naive student is molded into a functioning scientist.

Classes

The class requirements for first-year graduate students vary from department to department and from university to university. A past trend to reduce the number of required courses has resulted in some departments making few demands in this area. Still, today most departments do insist that students take a number of classes during their first, and usually second, years. These classes, some of which are left to the student's discretion, provide general and some specific background knowledge of a discipline so that research can be carried out intelligently. (It is often possible to petition to substitute an unrequired course for a required one, or to drop a core requirement altogether if one's background warrants it. Petitioning to make exceptions in one's case concerning requirements or rules of all types is quite common in graduate school.) Some of the electives can be chosen (in consultation with the advisor) to provide the tools of the trade. Common electives for this purpose include computer courses, statistics or other mathematical courses, and techniques courses of various types. Some programs require all students to take a series of techniques courses to prepare them for work in a variety of disciplines in their field.

Some required graduate courses will likely be of the survey type, and will provide a broad overview of a field. These are likely to have relatively large numbers of enrolled students, coming from a variety of different departments. These courses are likely to be conducted in a somewhat formal manner, in the fashion many undergraduate classes are. They are also prone to be very similar to an undergraduate class in terms of testing and perhaps even content (indeed, upper-level undergraduates may even be enrolled). They may, but need not be, more difficult than many undergraduate classes.

It is those courses characterized by small classes that are actually a hallmark of graduate school. These are attended by two to fifteen students, who commonly sit casually around a table instead of at separate desks, a set-up that makes conversation easier and that attempts to reduce the professor-student disparity. As you may expect, these courses (often called advanced seminars) have less structure and involve closer student–professor interactions, than do the courses with more students. The professor does not formally "lecture," but rather chats about topics and "exchanges ideas" with the pupils, who participate freely and ask questions openly. Students may be asked occasionally to present a topic and lead the discussion, and although this may sound intimidating, the atmosphere is really very easygoing. The key to making a good presentation is to make it in a *conversational* manner. These classes emphasize logic and reasoning (versus memorization), are less broad in scope than more formal classes, and are usually limited to selected topics that lie within the expertise of the professor. It is here that you get the real up-to-date lowdown on a subject: what is known and especially what is not known, who the major players are and what they are like, the problems in the field, etc. It is not surprising, then, that students start to feel like colleagues of the professor while attending such classes. While some of these courses use midterm and final exams as the basis for grading, it is also common for a paper (students write *many* of these), an oral presentation, or mere attendance to be used in lieu of an exam; in the physical sciences, however, problem sets are usually the basis for the grade.

It is easy for students fresh out of college to assume that graduate school courses are to be viewed in the same light as were the courses of their undergraduate years. They may assume, in particular, that class grades are very important for success, and that grades are a major factor influencing who gets the best postdoctoral or other job positions. This is not the case. Although some programs over-accept first-year students and then weed out "the poorer ones" based on grades, in most programs, "merely" *passing* courses is sufficient. This is not as easy as it sounds, however, as a passing grade in graduate school is typically a B, and some of the courses are difficult; two or three Cs on your record may be enough to get you booted. However, as stated above, some of the classes will be of the seminar type where a small talk, a paper, or just your presence is required, and where grading tends to be lax (or based on a pass-fail system). It is not necessary to obsess over grades; after you successfully complete a course, *it is unlikely that anyone will ever ask what grade you received.* While this may not hold true for eligibility for certain student scholarships, fellowships, or grants, it is almost always true for future employment: candidates for postdoctoral, faculty, and industrial positions are rarely asked to present their transcripts (although a transcript may be needed as proof of the degree). We do not mean to imply that classes are unimportant. Of course they are important, and they should be taken seriously, as there is a core body of knowledge that you have to conquer in order to consider yourself educated in your field. However, graduate school is about research, and research is where the emphasis should be.

◆   There were *lots* of homework assignments, and exams with *hard problems.* B was passing. (Students *averaged* a B in their grades—there was no grade inflation.) Some employers asked for grades, but I really don't think they are very important if you have papers. (Ph.D., Physics, University of Illinois)

No future employer is going to be horrified by your relatively mediocre grades (i.e., Bs) if you have a publication record that is excellent. We suggest only that you keep the proper perspective, and act accordingly.

◆   One of the graduate students in my lab would stay up all night studying for exams, and would be so exhausted after taking them, that she would go home to sleep. I, on the other hand, didn't take exams so seriously, and would return immediately to the lab to work. She would yell at me that I didn't have the right attitude, but it was I who got out a lot of publications and was the favored graduate student in the lab. (Ph.D., Psychology, biopsychology research emphasis, University of Virginia)

♦ Grades were of almost no importance in grad school. Almost no one ever received less than the "gentleman's B"—that was the minimum acceptable grade for grad students. Entering grad students at my institution were rather competitive and almost everyone tried hard to get good grades, but it was quickly understood that excellence in research was all that really mattered. The few formal courses I took were very tough compared with anything I'd had as an undergraduate, even though I had taken several of the first-year graduate courses offered at my undergraduate institution. During my first month or two of graduate school I became painfully aware of how poor my background was compared with the students who came from, say, Caltech. But the differences in preparation pretty well evened out by the end of the first quarter. (Ph.D., Chemistry, University of California, Berkeley)

♦ In grad school, they will kick you out if you get a few Cs. There is a lot more pressure on you in this sense in a Ph.D. program than in a med school. In med school, if you flunk a course, they may make you retake the course or try to figure out why it was difficult for you and how they can help you; they may even get a tutor to help you.

I think grad courses take more individual work. In college you have people around to help you. Grad school is a lot more isolating. From my experience, grad students don't work together. In med school, everyone works together in teams. No one said you had to, but we did. (M.D./Ph.D. student, Neurosciences, University of California, Los Angeles, School of Medicine)

Journal Clubs

A handy way both to help keep up with the scientific literature (a decidedly difficult thing to do), and to teach students how to critically evaluate that literature, is the activity known as the "journal club." Students, postdocs, and professors gather together at a set time (often lunch time, sandwiches welcome) to discuss a (usually recent) journal article in their field. Attendance at journal club may or may not be required but, either way, students had better show up most of the time. This is one of those activities where students and professors can participate equally, where everyone's input is welcome, and where all involved, no matter what their level, can learn something. It is also a good opportunity to interact with faculty that one does not see every day. The particular journal article(s) under consideration is announced beforehand, and attendees are expected to have read the article before the club assembles. The paper is then reviewed (usually by that week's presenter), and its merits are evaluated as the employed premise (Is it valid?), methods (Are they appropriate and performed correctly?), results (Were they analyzed appropriately?), and conclusions (Are they justified by the data?) are critically assessed. All can join in this critique, and many usually do, as taking potshots at one's scientific colleagues and competitors is part of the game. Being able to critically judge a paper takes much knowledge and years of research experience (even experienced scientists have great trouble evaluating papers outside their area), so new students should not feel insecure about a paucity of comments, and they should not feel obligated to speak. They will learn a certain amount just by listening to others praise the good papers and debunk the bad. For new students, observing this process can be an enlightening experience, and they will realize, perhaps for the first time, that even poor research can manage to gain admission into prestigious journals, and that even quality laboratories can sometimes publish scientific junk.

The journal club is not the most popular activity around. Many people find it boring or incomprehensible. Truthfully, it rarely is a thrill a minute, unless you have a particular interest in the paper being evaluated. But it is a good way to learn some things, and to get acquainted with papers that you might not read on your own.

♦   Everybody is bored, always bored. We try to help by ordering pizza, etc. (Graduate student, university/department withheld upon request)

♦   The only journal clubs that I have learned from are the specific ones, the ones that you are not required to take and that are in your area of interest. The required journal clubs—well, they should teach students how to expertly read papers before they make them take journal clubs. (M.D./Ph.D. student, university/department withheld upon request)

Lab Meetings

Many labs hold periodic meetings at which their grad students and postdocs discuss the progress of their individual research projects (general laboratory issues are also brought up). These meetings allow the lab head to direct, and keep abreast of, the work of the lab's members. Participants informally review the experiments they are engaged in, the results they have gotten, and the problems they may be having; the lab head and other lab members offer suggestions and perhaps criticism. Problems with experimental design, methods, data analyses, and conclusions are often revealed as the group discussion proceeds. These are not mean-spirited criticisms, and students should not feel downhearted because of them: lab meetings are a major vehicle for educating students about the careful thinking that goes on behind experimentation, and the difficulties and intricacies of the research process. In a way, lab meetings are like journal clubs, except that the research discussed is ongoing and close to home; more importantly, lab meetings try to catch any inherent problems with the experiments *before* they end up in print. Meetings are especially useful for students that are stuck at some tricky spot in their research, and need help from the group to get them through the problem. In many labs, meetings are held every week, and participants discuss their work of the week past. In some labs, everyone presents at each meeting; in others, one person presents one week, and another presents the next. Presenters can use slides, the blackboard, overhead transparencies, or handouts to present data and other information in the form of sketches, tables, graphs, etc. Although the general environment of lab meetings is casual and the presentations conversational (after all, the only people present are those that you work with every day), presenters should carefully plan out what they are going to say, and be well prepared.

♦   I went to some lab meetings as an undergrad and found them intimidating (I didn't have the background). Now I know the meetings are to work out your problems—"Hey, my stuff isn't working!"—then we come together as a group and try to figure something out. (Graduate student, Psychology, biopsychology research emphasis, University of Washington)

♦   I had to learn that science really IS in the details. (Ph.D., Physics, Princeton University)

Seminars

"Seminar" is a synonym for a lecture (the term is also used to describe a type of class, but that is not how it is being used here), and at a good university, there is always an interesting seminar being given somewhere on campus. The speaker may be local (a university professor, grad student, or postdoc), or a visiting academic from another school. Your attendance at a seminar may be suggested, but is not usually required (unless you are receiving credits for it). It is considered a courtesy to the speaker, however, if you do attend. When graduate students give seminars, it is usually in front of the department, or the students in the department, only. Even though giving a talk can be frightening at first, presenting seminars is good for students and is an essential part of the training process: by presenting their own research, students learn how to communicate their science, something they are going to be doing for the rest of their careers, something they need to do just to get a job. Scientists rise in the ranks only by publicizing themselves and their work, and publishing papers is just not enough. They must orally present their research at professional meetings and obtain invitations from universities for guest lectures, all so they can be seen, gain recognition, and interact with the well known and powerful. This is how they become players, and the more recognized they become, the more talks they have to give in order to stay on top. So, if there is no official means in their department by which students can practice the art of giving talks, they should seriously consider starting their own (unofficial) student seminar series. With a nonthreatening audience of fellow students, mastering fears and learning techniques can be accomplished with minimal anxiety.

It is considered especially prestigious for a university to be able to attract speakers from other universities who are leaders in their field. Do not miss out on the chance to hear such researchers discuss their work, even if it is outside of your immediate area. Guest speakers are typically invited because their work is unusually interesting and good. It is important that students are exposed to the research and minds of such scientists. If the speaker is someone in your field, you will likely find the room packed with colleagues from your department (grad students, postdocs, and professors), and there will be a sense of excitement in the air. Some of the lecturers, for all their experience and scientific skills, will be lousy public speakers. They will stammer, they will exhibit annoying mannerisms, their voices will not project, they will not look at the audience, their slides will be poorly designed and crowded with too much information, they will pace constantly, they will read from their notes, or their talk will be too long, poorly organized, obviously memorized, or dull. You will ponder how they became so well known, and rightly so. One wonders if such people would rise so high in today's competitive job market.

Other lecturers that you hear will be excellent public speakers, and their talks will be exciting and well organized. They will "tell a story." Clearly and simply they will relate why and how they did their work, and what the results were and mean. They will avoid superfluous material that will clutter the talk, and they will summarize their findings with emphasis. They will intersperse humor in their talk, and will pause often to let their points sink in. They will exude warmth, confidence, and calmness. They will field questions successfully and will know the techniques for getting out of tight spots. We mention all of this because you should watch speakers carefully so that you can learn from them. Of course, the speaker does not have to be a guest lecturer. Your own department will have excellent and poor public speakers for you to contrast. Emulate the characteristics of the good ones, the ones you enjoy listening to, the ones whose talk you grasp, and avoid the characteristics of the bad ones. Good speakers were not born that way; they learned how to be so, and so can you.

Often, time is specifically set aside for speakers to meet or go to lunch with students after their talks. Seeing these people "up close" and chatting with them about their research or yours are important experiences. You may find the prospect intimidating, but they are only people and most enjoy being surrounded by enthusiastic students (besides, it is a great chance to network). Many students have expressed the happiness that comes from having their ideas taken seriously by respected scientists in their field. Being treated as a colleague is a real high!

♦   After studying someone's work for several years, it is always a pleasure to have the opportunity to meet, and even collaborate, with them. (Ph.D., Computer Science, University of Southern California)

♦   It was wonderful when a senior researcher visited and he took me seriously and I found I could hold my own. (Ph.D., Psychology, biopsychology research emphasis, Dalhousie University)

♦   The most delightful event in graduate school is to have lunch with a famous scholar when he/she visits the school to give a talk. Having a personal or academic conversation with a person I've previously known only through a book really makes me feel proud and inspired. (Graduate student, Psychology, biopsychology research emphasis, University of Massachusetts, Amherst)

Finally, you will see something at seminars that will remind you that you are no longer an undergraduate. No matter how famous the speaker is, you will always see a few people get up during the talk, and quietly leave the room, presumably to continue with some ongoing experiment. In graduate school and academia in general, research always takes precedence.

5

The Absent Professor

In *theory*, graduate education is modeled after the classical apprenticeship program. That is, a student is envisioned as laboring daily under the guidance of a "master," a respected scientist who has acquired, over years of study and experimentation, the knowledge and technique necessary to perform sophisticated, state-of-the-art research. The student "protégé" works closely with the esteemed mentor in the laboratory, like a wide-eyed apprentice attending to a Michelangelo. The scientist imparts valuable knowledge and skills. The student observes, practices, and applies the mentor's methods, is coached and counseled at each step, and eventually is molded into a functioning scientist. This is a time honored means of ensuring the passage of craft and expertise to future generations.

In reality, however, the system does not work this way.

The practice of science nowadays is a complicated affair. As a result, contrary to what many beginning students expect, most professors who made their reputations by original work earlier on in their careers now have little or no time to spend in the laboratory. They are simply too busy. Surely one of the most time-consuming activities professors are engaged in involves the never-ending and arduous quest for research money. To keep their labs up and running, scientists need funds (commonly hundreds of thousands of dollars per year) to pay for supplies, equipment, animals, and salaries of graduate students and postdocs (and even portions of their own salaries). Funding is their most basic concern. A naive person may believe that the university itself supplies the money required for the research conducted within its doors. Except perhaps for some minimal start-up funds, this is not true. Most of the money is obtained from government (or, less commonly, private) agencies, and it is awarded infrequently. Scientists apply for these limited funds by submitting grant proposals; these are painstakingly detailed descriptions of the research that the scientists propose to do if money is indeed granted. Much of a professor's energy is devoted to preparing and writing these proposals: the background journal articles must be read and reviewed, the lengthy and detailed research proposal must be stated clearly, concisely, and logically, and a case must be made that the particular lab in question has the experience and know-how to correctly perform the experiments proposed. The percentage of grants that are funded has diminished greatly over the years, and more than one grant may be necessary to support the work of a single lab. Since new grant proposals or grant renewal proposals must be submitted periodically, for most academic scientists, the long and arduous task of grant-writing (not to mention the preparing of yearly progress reports) is seemingly ever-present, and takes priority over many other activities.

In addition to writing grant proposals, professors are also involved with the preparation of journal articles and books, teaching responsibilities, meetings, administrative duties, lab management issues, grant review "study-sections," and out-of-town lecturing. They also must think up new research ideas for lab members to work on. Theirs is not the world of the ivory tower. They are very much in touch with the practical realities of their profession. With so much to do, it is unlikely that most professors have found the time to master all the latest laboratory techniques, and they may not be totally caught up with the scientific literature either!

As a result of their arduous schedules, professors are likely to be in their offices working diligently on their computers, in some other part of the university at a committee meeting, at a grant review at the NIH, or lecturing in Germany. They are likely too burdened to personally continue with hands-on research. There are exceptions to this picture—some profs just refuse to give up bench work, and somehow manage to squeeze in experiments between other duties. But such professors are not particularly common (especially in biologically oriented laboratories), and students will have to search them out. Many new students who looked forward to close, perhaps daily contact with the renowned persons they carefully selected to work under, are stunned to find out that they sometimes have such little access to these professors. In fact, during especially busy periods, they may see their mentors only rarely, and they almost certainly are unlikely to see them in the lab.

> ♦   I started research in the lab of a brand new assistant professor, fresh out of his postdoctoral work. He was only in his 20s, but already he had given up working in the lab. He was just too busy writing his first grant, teaching, serving on committees, etc. He didn't seem to mind though. I felt that he liked the idea of being in control of a lab, and being so powerful that he didn't have to physically be there in order to keep it running. (Ph.D., Behavioral Sciences, University of Chicago)

The question then is, in that majority of labs in which the professor is "absent," *who* is doing the research that is published in the journals, research that maintains the lab's reputation so that old grants are renewed or new ones funded? *Who* is spending long hours reviewing the scientific literature, designing protocols, contemplating new equipment to purchase and new experiments to try? *Who* is learning the latest techniques, working out the kinks, perfecting the methodologies? *Who* is working into the night, gathering the data, making the latest discoveries, writing the papers, maintaining the professor's honorable standing in the eyes of the scientific community?

The graduate students and postdocs, of course!

Yes, graduate students, you are needed. If the status quo in academia is to continue, without you and the postdocs (who are not found in every lab), it could be that much research would slow down or stop cold. While good advisors are well informed about what goes on in their labs, thoroughly understand the science involved, and have the perspective necessary to oversee, interpret, and steer the research, you are likely the ones running the experiments, informing the advisor what works and what does not. While the professors are out there making themselves visible, networking and giving talks so that the money that runs the labs keeps pouring in, just remember that it is likely *your* work that they are talking about. And their labs are generally for you to work in; some professors have little idea how to use the state-of-the-art equipment on their lab benches. (Keep in mind though, that professors have their pick of students; when a number of students want to work in a lab that has limitations on how many it can hold, only those with the most promise will be selected.)

Disgruntled graduate students should realize that they are not the only ones to find fault with the way things are. Many professors wish they could return to the laboratory and perform the work that they were actually trained to do. They, too, are victims of the system.

♦ I know my advisor doesn't know how to do some of the techniques used in the lab. As projects evolve and new techniques come in, professors are usually too busy to learn them. I remember one professor who had to do a technique that he had last done when he was a grad student, and he had to come to a grad student in the lab for advice. Professors are just so busy writing and rewriting grants. (Graduate student, Neuroscience, University of Rochester)

♦ My advisor *did* work in the laboratory. I don't know if this is common, but I know some other professors in my department (physics) did the same. (Ph.D., Physics, Boston University)

♦ There is a range of technical competence when it comes to advisors. My advisor doesn't usually work in the lab, but knows most of the techniques used there, has done almost all of them, and can do them, although maybe not as efficiently as the technicians can. (Graduate student, Neurobiology and Anatomy, University of Rochester School of Medicine)

♦ Perhaps too many new graduate students expect that their advisors will actually spend much time doing research and teaching, rather than leaving this work (while still receiving credit for it) primarily to graduate students and postdocs. Perhaps too many students do not realize that one reason for their recruitment was to obtain cheap labor. Only a small portion of prospective graduate students realizes that they will be helping their advisors obtain grants, and thus advancing their advisors' careers. So, why not write a book to explain all these things to new students, and therefore prevent some of the confusion and unmet expectations that might strain student–advisor relationships and therefore hinder progress? (Ph.D., Biochemistry, Molecular and Cell Biology, Cornell University)

Indeed, the present system is surprising, especially to those who arrive at graduate school expecting a very different arrangement. To be sure, some students like working alone, and prefer not to have their advisor ever-present—but this may not be what they were anticipating when they arrived. Still, the relationship

between professors and students is symbiotic. Students learn to be
independent investigators by figuring things out on their own; the
genuine excitement and sense of pride stemming from that accomplishment is not to be downplayed. Also, students further
their own careers by linking their names on publications with the
names and reputations of their professors. Furthermore, while students work hard, they are also paid for what they do. In what other
situation are students *paid* to be students?

If the current situation is to change, it will require major
transformations at many levels, including the way research is
funded. Until that time, the way labs are run and the critical role
that students play are likely to remain fairly static. Yet, even
though the present arrangement is confusing and imperfect, students do learn their craft. In fact, they usually learn it quite well,
and by a variety of means, as the next chapter will show.

♦   I did an undergraduate summer project where I was working
with a grad student in a lab. I saw that the advisor wasn't around
for the student while he was doing his work, so, based on that experience, I was prepared when I started my own studies not to see
an advisor very often. I knew that an advisor was someone you report to, a person who shows up once in a while to see how things
are going. (Ph.D., Physiology, University of Wisconsin)

6

How You Learn

If professors are commonly absent from the lab bench, who then teaches the graduate students?

Ask any new Ph.D. who taught her how to be a scientist, and she will have to stop and think; the answer is not immediately obvious, and no one person really was responsible. There are not any courses on how to be a scientist, and no one takes you aside and explains it all to you. Graduate students pick up their practical and conceptual knowledge of science similarly to the way children pick up their knowledge of the culture in which they are brought up: they observe others, they imitate others, they get some guidance from authority figures, they read, they speak to friends and acquaintances, they are criticized when they perform inappropriately, they learn from their mistakes, they are rewarded when they perform appropriately. Interestingly, acquiring knowledge by such experiential methods (versus being systematically taught), is considered by educators and psychologists to be the optimal way to learn.

◆   I didn't see my advisor in the lab (thank God!). I knew some of the techniques from being in industry and the rest I learned from other graduate students and postdocs. My advisor was always accessible if I needed him. I pretty much only went to him when I was stuck and he would give me suggestions, etc., on what to do next and he helped me analyze confusing data.

I'm not quite sure how I learned to be a scientist and think like one. I think it happened spontaneously as I spent more time in lab, on campus and at conferences. (Ph.D., Chemistry, University of Rochester)

Modes of Learning

Observing Others

The prospect of being a "productive" scientist in the lab (or in the field) right from the start can be intimidating. How should graduate students proceed?

It is fine and proper to spend time when you first arrive merely observing others. In a productive and active laboratory, there will be a number of persons from whom you can learn: advanced graduate students, postdocs (who are also "in training"), and technicians; in some labs, there may also be a visiting professor from another university who is there to learn a particular technique. Most of these people will be happy to let you watch them as they work, and will gladly answer questions that you may have. Each is likely working on a different project, and may be using some different techniques. You may be assigned to one particular researcher (or to a technician) and be expected to learn the technique, or use the piece of equipment, they are using. Since you were good enough to get accepted to the graduate program, *you are expected to be a relatively fast learner.* Very soon you will be expected to do the procedure yourself.

It is probable that you will learn by observing others many, many times during your years as a student. It is understood in academia that those who know have some obligation to teach those who do not, and spontaneous, on-the-spot instruction on "this" or "that" is common. The ones teaching or giving advice will usually be the postdocs or your fellow graduate students (partly because it is less intimidating for one to ask advice from one's peers, partly because students/postdocs are often expected to do the teaching of lab skills, and sometimes are the only ones capable of doing it). Students who work out a method or technique during their research years may be held responsible for passing it on to others. It becomes their duty to see that the knowledge that they gained is not lost from the lab when the time comes for them to get their degrees and leave. (Unfortunately, this passage of information from student to student is one of the ways that lab heads get distanced from the workings of their own lab; they often are left out of the "loop.")

♦   My advisor was chairman of the department, and really had very little opportunity to be in the lab. However, he turned me over to four very capable men who *were* in the lab either full time or much of the time. All four were helpful, readily accessible, and gave freely of their time. I learned lab techniques from them and they helped me when I was struggling to work out new ones of my own. (Ph.D., Zoology, University of California, Los Angeles)

It will not be only through obvious training sessions that you will learn things from other grad students or postdocs. You will overhear them talking to each other and to professors, and you will pick up useful bits of information here and there. Also, "shooting the breeze" with fellow students/postdocs, perhaps the most casual and seemingly purposeless activity that students do, ultimately turns out to be, when looking back over the years, one of the most instructional and meaningful. These nonchalant, feet-up-on-the-desk chats—what the data mean, where the research should go, who is right and who is wrong—serve multiple purposes. For one thing, students can learn as much science listening to advanced students/postdocs as they can from a formal lecturer, although the information may come out in bits and pieces. Practical, conceptual, and theoretical information is routinely passed on this way. Furthermore, new graduate students feel less intimidated asking basic ("stupid") questions of advanced students/postdocs than of professors; thus, student-to-student/postdoc communication is often the main medium of transfer of some very fundamental information.

♦ I talk to the other people in my lab all the time. Techs have been great for learning techniques, but postdocs have saved my life in a lot of ways—how to FedEx and get my grant in—little tiny things that I didn't know how to do. The postdoc is the first person I go to to ask questions and interpret my data. I just feel much closer to the postdocs and more comfortable asking them all the "dumb" questions. They are much closer to the graduate level and they remember the experience. I go to my advisor more now than I did, but it is hard to ask the dumb questions of your advisor— you feel stupid. (Graduate student, Genetics and Cell Biology, University of Minnesota)

By talking amongst themselves, students also gain confidence. They become aware that each of them has distinct talents, and can offer unique insights to a problem. They share their ideas and wild theories with each other more freely than with the professor, and they get excited over each other's thoughts. From this positive feedback, they learn to respect their powers of reasoning, and their worth as budding scientists. They also share anxieties and insecurities, and as they become aware that they are not the only ones with apprehensions, they are reassured and gain the confidence to "go on." Students also trade observations on the virtues and flaws of the lab head. Hearing about the positive qualities of an advisor assures new students that the advisor will be there for them when needed, and/or that the scientific advice they receive is good. Learning about the negative qualities may result in a more lighthearted acceptance ("don't take it personally") of the advisor's shortcomings, and may make existing with this important, but imperfect, person easier to take; more advanced students may also be able to shed light on how to deal with the professor constructively.

Studenthood is a process these nascent scientists go through together and, because of all they share and all they learn from each other, a subterranean community of student-scholars evolves, where friendships are developed and frequently maintained, providing a framework for future networking. This is why if you attend any final oral exam or pick up any dissertation you are guaranteed to observe acknowledgment and sincere thanks given to fellow students for their role in the educational process and their part in the success story that that exam or dissertation represents.

♦   I learned how to think like a scientist mostly by interacting with other people. More specifically, this means being an active part of the lab, and the department, in general. Making sure to attend departmental seminars (even those not in your field), journal clubs, and thesis seminars are all ways to learn new things and, importantly, provide the chance to discuss these things with others. I always learned a lot by discussing seminars with friends afterward, or by giving a "recap" to a labmate who did not attend. And it goes without saying that being an enthusiastic lab member is important for this learning process. To me, this means chatting with others about their work, bouncing your ideas off them, and participating actively in lab meetings. Not only are these important ways to learn to be a scientist, but these interactions make up my best and most exciting memories of graduate school. (Ph.D., Genetics, Yale University)

Lab Meetings and Journal Clubs

Every man has a right to utter what he thinks truth, and
every other man has a right to knock him down for it.
                                                    —Samuel Johnson

As discussed in Chapter 4, lab meetings provide a means of educating students about the process of doing research. Very importantly, they are also a chance for students to interact directly with their advisors, and to absorb some of the best these mentors have to offer. Lab meetings teach thinking—thinking about theory, techniques/methods, analyses, meanings of results, and conclusions. By listening to the discussions going on at lab meetings, students learn how to formulate hypotheses from data, how to dissect hypotheses to reveal what they really mean and can or cannot predict, and how to design experiments that specifically test hypotheses. They also learn how to pick apart the constituent elements of experiments (considering all minute details and their possible effects), so that they can judge what the limitations of their techniques are, what problems they might expect, what they should and should not attempt, what their results will and will not mean, and what they can and cannot justifiably conclude.

This process of detailed dissection, of putting knife to theory and technique, is performed first when an experiment is still on paper; it is used to contemplate experimental design, methods, data analyses, and interpretation of conceivable results (the ability to anticipate results before an experiment is performed, and to interpret what these results will or will not mean, is critical for properly designing an experiment). The dissection is performed again after the experiment is actually carried out, to see if alternative explanations of the data are possible, if unconsidered factors managed to sneak into the experiment, and if it is necessary to repeat the experiment with more controls. These are analyses with intent to undermine, to destroy one's own work if necessary, and they serve to reveal problems and errors but also certitudes and truths. Serious application of this process earns a scientist a reputation as a (self-)critical, careful thinker. A student can watch this process being employed at a lab meeting for the purpose of analyzing a lab member's research, or at journal club for the purpose of evaluating a published article. Being witness to this process is a major way that a young scientist learns to judge and evaluate a piece of scientific work; critical appraisal by dissection, with its compulsive and obsessive search for "holes" (faulty thinking, errors in design or methods, alternative explanations of the data, etc.), becomes internally incorporated, and is the way scientists keep tabs on themselves as well as on others throughout their careers. It is the way young researchers grow to trust their own opinions, think for themselves, and be skeptical of, and even challenge, authorities. You will find it to be a repeated theme, and it is surely the most important research strategy that a student learns.

♦   Being able to critically examine ideas (my own included) came from my interactions with other scientists. (In some cases, after working with some particular people in science, it has been a learning experience of what *not* to do.) (Ph.D., Biology, University of California, Los Angeles)

♦   Though it is difficult for me to reconstruct how I learned to think like a scientist, aside from emulating those whose ideas I respected, I do remember when I first realized I thought like a scientist. A distinguished scientist was visiting our department to give a seminar. As usual, interested faculty and graduate students gathered afterwards at a local restaurant for drinks and dinner. We were all discussing the seminar and the speaker continued to expound his pet theories. I was not convinced and remember stating somewhat vehemently, "I don't believe that." Afterward I reflected, Who was I, a mere graduate student, to challenge this senior, very distinguished visitor? It was then that I recognized that I understood that theories live or die on the altar of the data. That ideas advanced by anybody, no matter how senior or how distinguished, could be challenged by anybody else. In this way science is very egalitarian and very interactive. (Ph.D., Psychology, biopsychology research emphasis, University of California, Riverside)

Private Meetings with Your Advisor

Meeting privately with your advisor in his office (something *you* should initiate, and arrive prepared for), or even having little chats with him at the bench, affords a wonderful opportunity to observe the workings of a scientist's mind "close-up and personal." It is perhaps the best way to tap into the wonderful educational resource that your mentor represents. You will learn how to think about your work, how to view it from a larger perspective and get the "big picture," and how to come up with new tactics when things are not working out. Private meetings can cover topics very similar to those discussed at lab meetings, or they can center around more theoretical issues. If you are lucky, your advisor may even wax philosophical, and let you in on his personal philosophy about science and research.

♦   My graduate advisor was a rising young assistant professor who earned tenure during my graduate career. He was quite accessible to his students and occasionally worked in the lab himself, but most of the details about how to do lab work were taught to me by the half dozen more senior students in the group. My advisor's main functions were to help me see my work in a larger context, suggest new ideas to try when I got stuck, and, particularly, provide encouragement when I got down or frustrated. (Ph.D., Chemistry, University of California, Berkeley)

Being able to share in the joys (and even the sorrows) of the research process with one's advisor is exciting. Animated discussions between professor and student about work in progress (a meeting of minds that spurs creativity) are very stimulating, and knowing that your advisor eagerly awaits your results makes your work interesting. The fun of celebrating important findings together adds to the allure.

Classes and Seminars

Obviously, students learn much of the fact and theory of their science by attending lectures in the form of classes and seminars. This forms the backbone of their understanding, upon which they gradually build and develop other associations more specific to their dissertation; the nitty-gritty details of their research area are learned by other means, such as reading the scientific literature on their own, and hands-on experience.

Reading the Scientific Literature

The reading of journal articles is not only about picking up specific bits of information on a particular subject, although this may be its main function. Each article from a respected journal can be thought of as a tiny, compact lesson on how science is done. Through the literature, students develop a feel for how scientists think, do experiments, interpret results; they observe different laboratories thinking differently, contrasting theories developing or dying, and carved-in-stone "tenets" crumbling. They see how crucial it is to thoroughly research a topic before attempting an experiment, so that studies are not unwittingly duplicated, the state of knowledge is acquired, thoughts are clarified, and ideas are roused. They learn how science is communicated through writing, and how to read between the lines of a paper, distinguishing various nuances of meaning. It is no wonder, then, that students collect photocopies of journal articles like squirrels collect nuts. Electronic, online versions of many journals are also now available, and articles of interest can easily be downloaded. If you want to learn to be a scientist, read, read, *read.*

Going to Scientific Meetings

Attending a national or international scientific meeting in your field is a little like being able to step into the pages of a journal article and talk directly to the authors. Here, you can attend lectures by famous scientists and ask them questions. Or, you can go to the poster sessions and chat for awhile with the researchers who stand next to posters they prepared describing their latest experiments (those not yet in print). Meetings enhance formal and especially informal interactions between researchers, and much knowledge is spread around this way. You can obtain new ideas, find out why your experiment is not working, learn new techniques, or see what the competition is doing. Matters discussed at meetings are usually state-of-the-art, so the air is charged with excitement. If you want to get a glimpse now as to what tomorrow is going to be like in your field, attend a meeting.

While the concept is downplayed, professional meetings (as well as the lecture circuit) also provide researchers with an excuse to travel to new and interesting places. Being able to travel to exciting cities in the U.S. or other countries while someone else pays the bill is one of the wonderful perks of being a scientist. People always manage to squeeze in some sightseeing during their trips.

Self-Resourcefulness and Learning by Doing

While some of science is cookbook-like, the rest is anything but. There will be aspects of a student's research that prove challenging and/or problematic, and these will require particular effort to work through, even with suggestions from the advisor. Depending on the nature of the formidable or sticky areas, problem solving may require serious and creative investigations of the scientific literature, clever and original experiments in the laboratory, the resourceful designing of new equipment or technology, or combinations of all three. Advanced students, because of their greater experience, are better at this than beginners—just by performing routine research for a number of years, one gets a "feel" for how things operate. It may take hours or years to work out enigmas—if indeed they *are* ever worked out—but the effort is usually worth it. The investigative process, including defining the problem, determining how to go about solving it, and solving it, is a major self-learning situation and one that cannot be taught in any class. Students thus teach a lot of science to themselves.

Sometimes the results of experiments require students to teach themselves whole new areas of knowledge. You go where the research takes you and experimental results can lead one into totally new areas, as demonstrated in these two scenarios:

Barbara, a graduate student in immunology, is using knockout gene technology to study the functioning of lymphocytes in the absence of a particular gene product thought to be synthesized only in the immune system. Surprisingly, Barbara's experimental mice are noted to have obvious neurological deficits. This finding, suggesting that the gene may also be normally expressed in the nervous system, is interesting enough to pursue. Barbara plans to perform *in situ* hybridization, a technique that she will now have to learn, in order to see where in the nervous system the gene is expressed. She will also need to study books on neuroscience, so that she can identify the structures that she will be looking at.

Mike, a graduate student in applied physics, is part of a team working on the development of a new medical technology called optical coherence tomography, a system similar to ultrasound but using light instead of sound to create images. The scanner is built, and tested in the lab, but there is not yet any evidence that it will be useful for diagnosing medical conditions. It is suggested that Mike now collaborate with medical doctors and apply the technology to study a pathological condition of the retina. To do so, Mike must now learn about the microstructure of the retina, so that he can first correlate the elements of the image with anatomy, and then compare the images from patients to those from healthy individuals.

Writing Journal Articles, Grant Proposals, and the Dissertation

The painstaking task of creating a scientific manuscript of any sort (they are all very similar) is a lesson in manuscript formatting and phrasing, logic, critical analysis and dissection, attention to detail, and problem solving. To finally produce a satisfactory manuscript, you cannot get around these. Your attempts at writing a report, a journal article, the dissertation, or other document, will be carefully critiqued by your advisor, and the criticisms, although constructive, can be very painful. It is especially common for students to make leaps of faith in their writing, assuming that readers know what they know, and what they are thinking. Scientific writing is a critically important part of graduate school and the profession, and is discussed further in Chapter 14.

Making Mistakes and Having Mishaps

While it is not something that is much talked about or readily admitted, the fact is that graduate students do make major mistakes and do have accidents in the lab.

They think sloppily, design their research poorly, lose important data or forget to record it. They use the wrong reagents or leave out steps, rendering their expensive experiments useless. They forget to properly care for their instruments, leave the lab a mess, or break safety regulations. They drop and break instruments and equipment, and cost their advisors hundreds or thousands of dollars in losses.

At one time or another, such embarrassing incidents happen to all students, and they are forced to bear the critical words or disappointed looks of the lab head. Errors in thinking or practice are very common, even more so during a student's early years.

♦   Just today, I was running an ELISA with 14 plates of samples. I have been pipetting and pipetting for the last three days. Today I made a mistake and mixed up some samples. I was upset, but I put the screwed up plate aside and didn't use it. Later I realized that the plate that I put aside was not the one I thought, and was really the plate containing the *controls*. Because of that error, I lost one and a half days of work. (M.D. and Graduate student, Biology, University of Michigan)

♦   I started a hot project . . . the project was hot in more ways than one—in the rush to get early results, I got some serious burns on my hands from an accident in the lab. (Ph.D., Chemistry, MIT)

♦   I accidentally broke a $4,000 sequencing gel apparatus (this was the first and only time I had done anything that bad) and really felt awful about it all. (Graduate student, Virology, University of Alabama, Birmingham)

♦   The lab was running a long electrophysiology experiment, and after we finished recording, we injected a dye into the cell that we were recording from so that we knew which one it was. I was the one who later did the histology, and one of the sections that I was cutting just flipped off the slide and was lost—it turned out to be the cell we were recording from, and it made the experiment useless. (Ph.D., Biology, University of Texas)

◆   Another student in the lab was running a DNA gel, but using the power supply that I wanted for my protein gel. I asked if I could switch his to another, but he said no. I reasoned with him (I needed the timer function, and he didn't) and pleaded, and he finally said OK. So I switched his gel to another power supply. What I did not realize was that in the process I reversed his leads, and ran all his DNA off the top of the gel. I offered to rerun the gel for him, but he told me that the gel I ruined contained the last of his DNA. Oops. He was pretty mad at me. (Ph.D., Biology, Yale University)

Importantly, one learns from these painful moments. One learns how to behave in a lab, how to care for lab equipment, how to think and plan beforehand with care and rigor, how to record all that transpires, how to anticipate certain types of problems before they happen, and how to concentrate on what one is doing at all times. Hopefully, one learns not to make these same mistakes again. Even the seasoned investigator, however, makes errors, especially when attempting something new. Experienced researchers have learned to be able to anticipate certain problems and prepare for them, but some problems cannot be predicted. It is only when nothing is learned from a failed experiment that it truly can be considered a waste.

Constructive Criticism

Picture the arrival of an enthusiastic new student. He thoughtfully reads the papers that have come out of his new lab, formulates some neat experiments to try, and with great expectations, knocks on the door of his advisor to present his novel ideas. The professor is impressed with the initiative displayed by this new lab member, but as the student listens in disbelief, she gently explains why the proposed experiments could not and should not be carried out. Perhaps the hypotheses are too broad and ill defined, or the experiments hopelessly incomplete, and lacking controls. Perhaps the studies would require many years to complete, something a beginning student could not anticipate. Maybe there would be too many ways to interpret the results, making any conclusions from them meaningless.

The student leaves, dejected. He has had a tiny glimpse into a new world, where all thinking strives to be rigorous, exact, and scrutinizing. He has learned that he is going to have to develop a new way of looking, a more penetrating way of reasoning. The student has experienced a little constructive criticism, perhaps the most influential and broadly used method of teaching graduate students how to think and behave like scientists. It is guaranteed that a student will not get through graduate school unscathed: he will receive some type of constructive criticism shortly after arrival on campus and it will continue on, right up through the dissertation defense. But the comments are not always fault-finding; one also learns what is good and acceptable by critiques.

Constructive criticism or constructive praise comes in many forms:

1. Comments written on class papers or exams. ("You're jumping to conclusions here—it proves nothing of the sort"; "You would need a proton polarimeter with polarized and unpolarized electron beams"; "I like your idea for testing the hypothesis!"; "Content OK, but poor organization.")
2. Comments in the lab. ("Use proteinase K next time—it works much faster to strip proteins"; "That's a very dangerous way to handle the electrodes.")
3. Comments at lab meetings. ("You need to control for that possibility"; "You used the wrong statistics"; "There's another explanation for the data than the one you are giving"; "The major problem with your premise is that you have no way of knowing for sure if glucose use and blood flow fluctuate together.")
4. Comments from the lab head written on early drafts of a paper to be submitted for publication. ("You left out details of the incubation step"; "Use a review article to cover the first four references"; "Your results are *consistent* with that conclusion, but you can't claim you *proved* it.")
5. Comments from journal reviewers concerning papers submitted for publication. ("I do not recommend publication of this paper—no meaningful information can be garnered from sampling a known cyclical hormone at one point in time"; "The authors make a strong case for the involvement of nitric oxide in pituitary hormone release. This is

an important finding that is worthy of publication"; "There is no reason to include Figure 4 in the paper—the information is already in Table 2"; "The photos are too poor to reveal the bands clearly.")

6. Comments by the dissertation committee at the preliminary oral exam. ("You never mentioned what you are going to do if an animal does not exhibit the behavior at all—how are you going to include that in the statistics?" "There are going to be too few cells undergoing apoptosis to detect by electrophoresis. You are going to have to use the TUNEL technique instead"; "Because of what you are looking for, you are going to have to add a preliminary step removing the mitochondrial DNA.")

7. Comments by the dissertation committee at the final oral exam. ("I think you are going to have to redo that analysis using transformed data to make it meaningful"; "I'd be even more convinced if you also found magnetite and iron sulfate precipitates.")

Such constructive commentary results in the fine-tuning of a student's thought processes and technical skills. Through negative and positive feedback, a student's actions and judgment are molded until they approach an acceptable standard.

♦ People have told me that a scientist is someone who knows how to ask the right questions, i.e., formulate the right hypotheses so that one can run experiments and get some answers that are valuable and not worthless. I know that a lot of students coming in don't know how to formulate an hypothesis, and they don't know how to judge their results. Ask your mentor—is this a good hypothesis? Is there a way to test this? You have to ask for it. You have to learn how to ask the other grad students and fellows. It's all constructive criticism. That is how you learn—criticism and experience. It is all "job experience." (Graduate student, Microbiology and Immunology, University of Miami)

♦ Constructive criticism was very important in my development as a scientist. It was one of the most important things I learned—possibly because it was, and is, so difficult. (Ph.D., Zoology, University of California, Los Angeles)

♦ Constructive criticism is an important part of anyone's scientific development, but the older students in the research group were reasonably kind even if they occasionally liked to flaunt their superior knowledge, and my advisor, knowing that I tended to be my worst critic, was careful to put a positive spin on most of his criticisms to avoid discouraging me. (Ph.D., Chemistry, University of California, Berkeley)

These then are some of the many ways that graduate students learn their craft. Over the years, as they watch lab members and professors, and converse with them; listen to lecturers and seminar speakers; attend journal clubs, lab meetings, and professional meetings; work through obstacles in their research; develop a feel for their research area merely by working in it; read countless journal articles and write manuscripts; are challenged to think deeply and carefully by advisors, teachers, and committee members; learn from their mistakes; perform unsuitable work that brings discouraging consequences, and suitable work that is rewarded and reinforced—graduate students slowly but surely begin to start thinking and acting like scientists.

♦   When I started in the lab, I just watched the others. Then I asked more questions and tried the techniques myself. You watch how to mix up solutions, use the various meters, etc., and then you do it yourself. You learn everything from the other students. It's not a structured thing where you go to a class to learn. You learn to *think* like a scientist by watching too. You watch people in journal club, you watch people give seminars, you watch professors discussing their research. It's a long process but soon you are talking like them and walking like them. It's just a matter of time and exposure. (Ph.D., Biology, University of Chicago)

Graduate school is an emotional and intellectual "process," and the process and learning are synonymous; every decision, every success, every failure, every mistake, every struggle, every quandary, and every challenge is part of the process. Graduate school is an individualistic experience—no two students, even those from the same lab, learn the same information, skills, or techniques. It is hard to say when the naive student is transformed into a scientist. Analogously, can one pinpoint when a child starts thinking and acting like an adult?

♦   It seems like it would be easier if there was some organized structure, if someone would say, "People, this is what we expect, this is what you must do and learn," rather than having us try to figure it out. If you are a cook, you have to know how to make cakes, soups, breads, etc.; then you can say, "I am a cook; this is my profession." It is very intimidating for me to say "I am a scientist." I don't even know what it means to be a "scientist." There are people who are already there, and there are people like me. Nobody takes your hand and guides you through every step. You are thrown into everything. You have seminars, you go to classes, journal clubs, and lab meetings. My impression is that you are expected to somehow pull it all together, but nobody "teaches" you anything. (Graduate student, Neuroscience, University of Rochester)

The change happens unevenly, occurring faster in some domains than in others. But eventually, the pieces slowly come together: the technical skills (and the awareness of their limitations), the knowledge of the literature, the rigorous and critical thinking necessary for forming hypotheses and designing experiments, the insight gained from experience. The changes occur so gradually and are so multifactorial that the students themselves are likely unaware that a profound transformation is occurring that makes them more skillful at thinking and reasoning than those who have not received such training. They blossom into scientists rather incognizantly.

7

Deciding on Research Projects for Your Dissertation

The Ph.D. degree is awarded for conducting "original research." This requirement, researching something that no one has researched before, is less intimidating than it sounds. A small, logical step forward can constitute "original research." If it has been reported that the amygdala of the human brain is activated during fear responses (as examined by positron emission tomography), and you use the same technique to investigate whether the amygdala is activated by other emotions, such as happiness, this is "original research." Still, coming up with appropriate dissertation projects, whether they be theoretical, methodological/technological, and/or experimental, can be difficult and frustrating. Your projects will have to be narrow enough to be researched thoroughly, and the problems/issues being examined must be well defined.[1] The projects will have to fit in with the general scheme of the lab, and should not stray too drastically from the subjects presently being studied there. Ideally, they should build on the lab's past findings: explaining observations, studying discoveries in depth, exploring new leads. Your experiments, however, must stay within the lab's budget, which usually is dictated by the professor's grant. They

---

[1]"Defining a problem" is a decision-making task. A student who wants to study "the effect of stress on immune function" must specifically decide which stressor to use (loud noise? shock? maternal deprivation?) and what is meant by "immune function" (cytokine levels? lymphocyte numbers? ability to resist an infection?). If he/she decides on "loud noise," how many decibels are "defined" as loud? If "lymphocyte numbers" will be used to define immune function, which types of lymphocytes will be counted and from which tissue will they be collected?

also have to be in line with the technical sophistication of the lab; do not attempt a project that is beyond the capabilities of the lab unless you have lined up another laboratory that is familiar with the methodology of your needs and is willing to assist you.

Many professors will make it "easy" for you; they will assign some initial projects to you. These usually are experiments that they have been thinking about, perhaps ones explicitly laid out in a grant. For some students, this is the perfect way to go. The professor will be enthusiastic and happy to help (having a personal investment in your success), the methodology is available, and the research has a "history" in the lab, with many of the kinks or techniques already worked out. While doing these assigned experiments, you will learn valuable skills and gain a feel for research. Inevitably, the results gotten from these original studies will suggest further experiments to you and you will learn to think like a scientist, building on past findings. The projects will soon take on lives of their own, twisting and turning and evolving in ways unexpected, and you can rightfully take responsibility for interpreting and responding to each shift and transformation with creativity and thoughtfulness. As the project becomes more and more yours, your interest in it will grow.

♦   I didn't pick my dissertation topic. The professor assigned the project to me. It was a theoretical study—computer modeling. He *assigned* it, but its *development* was mine.

I did computer simulations of materials. This involved modeling, development, and computer display. My advisor and I would develop the models together. He would suggest the original idea, and I would develop it. I also picked up some things from the other grad student in the lab. I was doing an extension of what someone else had done before me—you do not usually start out of the blue. I used this earlier work as a guide. (Ph.D., Physics, Washington University)

♦   I got accepted to every graduate school I applied to. The professors got angry when I disagreed with them that working on a *professor's* project was something "independent." To me, there was something definitely not upfront about this, so I wisely chose not to get a Ph.D., and set up a laboratory in my garage. I have since received seven U.S. patents. (President, Biosource Inc., Worcester, MA; former graduate school applicant, University of California)

Often, the questions a scientist tries to answer are unpretentious ones. If the ultimate problem under investigation is very complex, one must chip away at it in small increments. While each individual advancement may seem insignificant, together these small parcels of knowledge can present a formidable picture of reality. You may find that working on a project addressing a modest issue is frustrating for you but, in fact, this is the way that much of science is done. We hope and trust, however, that you will keep the bigger picture in mind as you work and that you will strive not to substitute triviality for simplicity. (Some researchers actually prefer to continually work on modest and main stream issues because they know that these are the ones most likely to yield some results—in these times of funding shortages, job insecurity, and publish-or-perish mentality, it is often what works that ensures success, and papers on humble issues can certainly fill a curriculum vitae or "cv," the résumé of the academic world. This is an unfortunate state of affairs. At least occasionally attempting innovative, risky, and original projects that employ novel concepts, approaches, or methods is important for science's progress. Students, however, should not feel pressured to be so inspired).

Students deciding on dissertation projects should realize the very real possibility that they may not end up in academia. With jobs at universities growing harder and harder to get, many scientists are turning to industry, government, or other institutions for employment. Students that are already thinking seriously about such careers are wise to pick dissertation projects that involve topics or techniques applicable to these nontraditional jobs, as they may then have an easier time getting hired and will make a less stressful transition. This important issue is discussed in more detail in Chapter 16.

Strategies for Success

Most graduate students explore several avenues of research (sometimes for years) before finally settling on the topics that eventually make up their dissertation. Whether you design your own projects or "accept" those suggested by your professor, there are a number of points that you should consider before you proceed.

It is a wise idea to have at least two projects going on at the same time. Simultaneous studies increase your chance of success; that is, they increase your chance of finding a study that can be carried out to completion. As even relatively "safe" projects can prove disastrous over time for one reason or another, by pursuing more than one project you can stick with those that are most promising and, if necessary, drop the rest. This strategy can be especially useful to those that need daring and intellectually challenging research to satisfy creative drives, and thus are eager to attempt experiments that are theoretically or technically fraught with risk. While risky "long shots" are fun to do, and while the payoffs may be big, the odds for success are not good. It may take years to work out the methodology of an experiment, even if it ultimately must be abandoned, and no potential employer is going to be impressed when you relate how many hours you devoted to a project that did not pan out in the end. Thus a risky study should ideally be combined with a surer one to help guarantee success.

Long-term projects can also be considered "risky." They can take years of your valuable time and still may result in only one— or, even worse, zero—papers in the end. *Never agree to take on a long-term project unless you are also working on a few short-term projects at the same time.*

There is an additional advantage to having concurrent projects: they allow for more efficient time management. While you are waiting for the assay for project A to finish, you can analyze the data from project B, and while you are waiting for the rats from project C to habituate, you can start writing up the results of project B.

◆   It is very hot nowadays to determine the structure of proteins, but to do this you first need to purify the protein and grow crystals. This can take a long time and be very difficult. I tell my students not to try to determine a structure unless someone has already done the purification and grown crystals. Or, if they do want to attempt purifying proteins/growing crystals, I make sure they have two projects—the second project is a surer one. That way, they always have one project that they can fall back on if the other proves too hard to do. (Ph.D., Biophysics, University of Rochester)

◆   You should have 2–3 projects that CAN be done, projects for which the techniques are in the lab. Unless you want to spend 7–10 years in the lab, do not have more than 1–2 projects that require techniques that the lab doesn't have and that you have to work out. (Graduate student, Physiology and Neurobiology, University of Connecticut)

◆   If you tackle an extremely hard project, one that people have been working on for 100 years, unless you are very, very good, you probably won't get results. You will find this period to be very frustrating. You can take a crack at such a project, but it is important to know what you are getting into. (Ph.D., Physics, Case Western Reserve University)

◆ The most important guidelines for choosing a project in the physical sciences are:

1. Make sure that the project is well defined. There is a strong attraction, particularly in theoretical physics, towards difficult problems that have never been solved, big earth-shattering theories, and "fundamental" discoveries. Save it for when you get tenure. Do not waste years in frustration (as I have seen many times over). The point of the Ph.D. is not to "make your mark" in the field, but to establish a groundwork of skills that will make you a competent scientist.

2. Do not be discouraged by projects that appear to be too straightforward. Initially, you should feel like you could do the research in a couple of years. Once you get into it, have no doubts that your studies will take on a life of their own. (Ph.D., Physics, Princeton University)

◆ I had two related projects, either of which could if necessary have stood alone as a thesis project. My primary project was one that no one else in the group (or the institution) had worked on before, so I was the unquestioned expert, and while it was often intellectually very challenging, it was also fascinating and deeply satisfying. The other project was one that a student had worked on before me without complete success—and, as it turned out, I would not be completely successful either. I never enjoyed that one as much and often felt that I was spinning my wheels to little effect. (Ph.D., Chemistry, University of California, Berkeley)

The Importance of Publications

It is important that you realize that all the studies that you perform for your dissertation do not have to "work." That is, they do not have to result in findings that can be published. As long as the studies are carried out to completion, showing logical progression of ideas, logical interpretation of results, and careful technique, they are likely sufficient to be included in the dissertation. As long as you can show that you can think and perform like a scientist, you will eventually be awarded your degree, even without any publications. However, in order to increase your chances of future employment in this publish-or-perish business, you have to build and maintain an image of productivity—publishing your work should be a top priority that starts in graduate school and continues throughout your career. A publication tells the world that your work has been judged by outside reviewers and found deserving; your publication record is the official declaration of your "worth" and effectiveness, and *a good/consistent publication record is singularly the most important asset a scientist can have.* Your publications as a student do not have to be earth-shattering; good, solid pieces of work are fine. Do try to get your work published in the best journals in your area that will accept it; the standards of journals vary, and a publication in a well-respected journal carries the most weight.

Considering the importance of getting your work in print, you should be aware that, in general, scientific experiments that yield negative results (the data from control and experimental conditions are indistinguishable) are rarely accepted for publication. Negative experimental findings are frustrating, possibly detrimental to a career, and all too common. However, it is often possible to work on experiments that are publishable no matter what results are gotten! Surprised? This is a technique employed by many a sharp researcher, and you too should take advantage of it when you can. In essence, an experiment is specifically chosen because it has a very practical characteristic, i.e., whether or not the hypothesis being tested is accepted or rejected, the conclusion is still publishable, being a valuable contribution to the scientific database. Let us look at examples of such experiments:

1. Reports in the literature appear to show that Factor X is responsible for the tonic release of pituitary Substance Y into the bloodstream. You wonder if Factor X is the *only* factor controlling Substance Y release. A pharmacological agent that blocks the bioactivity of Factor X has become available. If you administer this agent, and then measure the amount of Substance Y in the blood, any result that you get would be potentially publishable. If blood levels of Substance Y go up, go down somewhat, or remain unchanged, it shows that factors other than X can or do control the release of Substance Y; if Substance Y levels become undetectable, the results are consistent with Factor X being the only releasing factor.
2. There is a clear cut T-cell/B-cell dichotomy in the immune systems of teleosts and amphibians, as well as mammals. You propose to investigate if the same is true for lampreys. If your exhaustive experiments are well designed and critically

executed/analyzed, your results should be publishable no matter what the outcome. If a dichotomy is observed, then lymphocyte heterogeneity evolved prior to the emergence of bony fish—an important bit of information; if a dichotomy is not found, the scientific community would similarly be interested in knowing that the phenomenon may not be specific to all vertebrates. Comparative studies can be very promising; just make sure you are studying a phenomenon that is biologically significant enough to warrant the investigation and that your papers have appropriate caveats.

3. There are two competing theories on an aspect of semiconductor nano-crystallite systems. You realize that under certain experimental conditions, each theory would predict a totally different outcome in terms of certain measurable parameters. No matter how the results go, in favor of one theory or the other, the experiment is publishable. An experiment that helps decide between competing theories is always a winner.

4. The Fas/Apo-1 receptor, which appears to be involved in programmed cell death (apoptosis), has not yet been fully characterized. Since this receptor is of great biological importance, analyses of its physiochemical properties are highly publishable. Such *characterization* projects may be the way to go if your skills and interests tend to be methodological.

◆   I worked for a professor who gave me lots of freedom to study what I liked, but little advice. For *four* years I performed experiment after experiment, each ending in dismal failure. I did not have a single finding to report. Finally, in desperation, I sought out a professor that I respected and asked for help. He was aghast at the experiments that I had been attempting. "Didn't anyone tell you," he asked, "the trick of only working on a sure thing?" (Ph.D., Biology, University of Michigan)

Another practical strategy for career-minded students is to avoid working for many years on a complicated project that will obviously yield only one paper in the end. The best project to attempt, if possible, is one that can be broken down into separate, publishable parts; that is, the project is composed of independently publishable but related studies that build on each other or, the results generated as one gathers necessary background information can stand by themselves and be published as individual papers. For example, say you are trying to determine the physical basis of a certain phenomenon. You first need to gather some data on the system and its dynamics so you know where, when, and how to focus your efforts. If these background studies will provide important information on the properties of the system that is general enough to be of interest to others, the data that you generate here should be publishable. Thus, you will make the most of your work and little will go to waste. There is an additional advantage to this tactic: if for some reason you abandon the main study, it will not be a total loss; you will still have one or more papers to list on your cv. (Publishing many small papers on a topic, versus one large, comprehensive, multiexperiment paper is a controversial choice, however, and a sign of the times. The former mode obviously "pads" the cv.)

Students that research "hot" topics or make use of "cutting edge" techniques have a significant advantage in terms of employability (which topics/techniques are currently in favor, however, can change quickly). In addition, if one's research area has clinical or commercial applicability, it has a good chance of long-term funding. Students who are lucky enough to have access to a *brand new* technology or methodology have a wonderful opportunity to become well known while still young. Make the most of your good fortune—ask new questions, explore areas never studied before. The invention of immunohistochemistry afforded such an opportunity to a number of neuroscientists, who quickly became famous applying the technique to identify the chemical systems of the brain. It is a lot easier to make the most of a new technology that is already perfected than to design and perfect it yourself. Being the one to create a new instrument or discover a new method may have huge payoffs, but results will be a long time in coming and success is not assured. Such a path is not particularly recommended unless you have significant, short-term projects going on at the same time.

There are certain types of projects or experiments that students should avoid. These are ones for which the results will not be clear-cut, easily decipherable, or predictable; they are also iffy when it comes to publishability. Experimental designs should always be closely scrutinized so that the possibility of obtaining ambiguous results from which conclusions cannot be drawn, can be foreseen and/or minimized. Your advisor should be able to help with this analysis. An example of an experiment for which the results are totally unpredictable is the "fishing expedition." Here, there is little in the literature to go by; one is just "fishing" for a find, lowering the bait to see if anything gets hooked. This is not usually the type of experiment that a graduate student should perform (unless it is quick, simple and inexpensive), as there is too great a chance that a finding will not be made. Let us use the "ultimate" fishing experiment as a simplified example:

The discovery of a fossil of a new order of Devonian fish is creating quite a stir in the paleontological literature. A beginning student proposes to look for fossils of this type in deposits in her state. Would that be wise of her? Probably not. Although the finding of such fossils would be significant, the student's chances of failure are too high. And what if she did not find any, even after years of searching—could she publish that? It is not likely; a negative result like this is not usually publishable. Perhaps she did not look in just the right area, or maybe she did not recognize the fossil when she saw it. An editor is never going to take the chance of publishing her conclusion that this order did not exist in the region just because the student could not find any evidence for it. This is a prime example of a result that is uninterpretable and therefore unpublishable.

Enthusiasm can work against some students. It is possible to have so many interests, so much curiosity, and so much zeal that the ability to focus on a single area of research becomes difficult. Students with these characteristics can bounce around from research subject to research subject, publishing a number of disconnected papers while never developing true expertise in any one area. This is unfortunate; for the sake of future employment, it would be much more advantageous if the work of these students had a definite theme. When it comes time for students to give talks on their research to potential employers, including postdoctoral supervisors, they should have *a story to tell:* they should be able to describe the background of their research, how their work fits in, how each individual experiment progressed logically from the one before, and how their results add to the understanding of the field. They do not want to be perceived as dabblers in science; they want to be recognized as experts in their area of concentration. After all, it is only by consistently publishing on a particular topic that one's name becomes recognized and associated with that topic.

◆   The primary problem for me was to learn to *focus.* I tried to go off in too many directions with my research, and thereby wasted a lot of time and energy. When this finally dawned on me, I managed to focus on what needed to be done and then did it! And yes, it was difficult for me—there were so many interesting projects I had to ignore in order to do it. (Ph.D., Zoology, University of California, Los Angeles)

♦ This grad student I knew was interested in a lot of things and was jumping around doing research in a lot of areas. I took him aside and said, "Hey, I used to be like you. It's great, but you have to decide if you want to stay in academia. If you do, you're going to have to focus and work steadily on one topic. People only read the papers in their own area. If you publish four papers in four different areas, then as far as the people in each of these areas are concerned, you've only published one paper. You'll have no name recognition." (Ph.D., Physics, Brown University)

In order to have a good story to tell, it is important that the topics that you are working on are meaningful ones. If you are determining a protein's crystal structure, the protein should have some significance; a protein implicated in cancer, for instance, would certainly be worthy of study. Determining the structure of a protein with no known biological function will not make a good dissertation project.

♦ A graduate student should always pick the smartest advisor possible—with the best "taste" and the sharpest mind. It's a waste to work on projects that sound exciting but, in fact, have very little content or possible scientific impact. (Ph.D., Physics, university withheld upon request)

For some students, the work that they do for their dissertation becomes the basis of their future careers. For others, the research that they started out doing is dropped after graduation and never picked up again; many scientists are doing research today that is very different from that which they did as students. Typically, those who change areas gain their new expertise during postdoctoral training; students commonly move into a new field by accepting a postdoctoral position in a laboratory with a focus different from that of their predoctoral lab. Those who switch their research area still later on in their careers face initial difficulties getting money for their projects. Grant committees must feel confident that an applicant's knowledge and skill is sufficient to correctly perform and interpret experiments proposed and, for people who have changed focus and have no publications in their new arena (and who may be self-taught in this area), the issue of competence is especially questionable. Until one has proven himself by publications, it may be necessary to collaborate on a grant with an accomplished researcher in the field; alternatively, spending a sabbatical working in the lab of such a researcher may be enough to put a committee's mind at ease.

The Scientist as a Specialist

A common misconception about graduate training is that it will produce scientists who are broadly and thoroughly educated in their subjects (i.e., the general fields [chemistry, biophysics, etc.] that are listed on their diplomas). This may have been somewhat possible a hundred years ago, before the explosion of scientific knowledge that marks the twentieth century, but the concept is truly absurd and unattainable today. Nowadays, students could not receive an exhaustive, general education in their fields if they were given a lifetime to try. Partly as a result, the scientists that emerge fresh from the modern graduate program are specialists. They are knowledgeable in their subdivision but true masters only in their specialty, that narrow (albeit complex enough to take years to study) domain that they focused on for their dissertation research. In this area, simply because of the narrowness of the focus, it can truly be said that the young scientists are, without a doubt, some of the world's experts. But their expertise does not extend too far beyond this realm. And while graduating students know a number of key concepts in other subdivisions of their major discipline, enough perhaps to teach some undergraduate or introductory graduate courses in these areas, there is really no great depth to their knowledge there.

> ◆   I did not expect to work on such a narrow range of research. I spent six years of my life studying two cells in the crayfish (exploring what second messenger system they utilized, the nature of their photopigment, the possibility of specialized structures as seen through an electron microscope, etc.). I ended up knowing a LOT about one little thing, instead of a lot about lots of things, as I thought would be the case. I had a cartoon taped up on my wall which summarized what grad school was turning me into. The cartoon was of a Ph.D. being introduced to someone as, "Dr. X, who is an expert on crocodiles," to which Dr. X replies: "You are too kind—my field is the crocodile eyelid." (Ph.D., Biology, University of Texas, Austin)

There are recent attempts to rectify this situation somewhat: departments are striving to produce students with a greater breadth of knowledge, a broader background of information. This is a wonderful goal, and one that should be strived for. Practically speaking, however, individuals can learn just so much outside of their working domains and still keep up with their own areas of research; there is simply too much to know. Once this is understood, students should be able to guiltlessly resign themselves to their relatively "limited" awareness. This should act to relieve them of some of the insecurities often felt at the beginning of graduate education; beginning students can easily feel overwhelmed sensing the enormous and diverse body of knowledge in their fields. A large scientific meeting, for instance, with its ocean of presented material, can readily target a student's vulnerabilities. All those posters, all those talks—how can people be expected to keep up with *all* the information and new developments out there? In fact, they are not expected to, and the sooner a student realizes this, the better.

♦ Scientific meetings were intimidating and overwhelming. I actually thought that someday I would be expected to understand *all* of that stuff! (Ph.D., Psychology, biopsychology research emphasis, Northwestern University)

♦ When I didn't have a thesis project (during my first few years), it was almost worthless for me to go to professional meetings just to see what was out there. I felt very insecure; I didn't know what I was doing. The Neuroscience meeting had 20,000 people, and there was a sea of thousands of posters each day. I understood 0.0001 percent of them. Even now, I do not understand most that are outside of my thesis area—I understand maybe one percent. Even of those posters that are in my area, there is only a small percent that I feel I understand well enough to ask questions of the presenter. (M.D./Ph.D. graduate student, Neuroscience, Brown University)

While students should seek to broaden their knowledge, and should keep broad goals, it is still comforting for them to realize that mastering all aspects of a subject area is not what is expected or demanded. Students very soon become totally absorbed in their specialties, and other areas of information, while not forgotten, tend to fade into the background.

♦ Advanced grad students are very focused. They have little time or energy to explore subjects outside of their dissertation topic. I find this distressing, but what can I do? You can't do everything. (Graduate student, Anatomy and Cell Biology, Vanderbilt University)

The subdivisions of every scientific field have undergone vastly divergent evolutions, and they are now so different and distinct from each other (in subject matter as well as in technology) as to be unrecognizably related. Because of this, there will always be some particular subdivision that will not match the tastes or talents of some particular student. Thus while some paleontology students are happy measuring in the (air-conditioned) laboratory the relative amounts of uranium and lead for zircon dating, others are thrilled to be digging up fossils of dinosaurs in the Mongolian desert; while some neuroscience students are staring at their computer screens, engrossed in their work on computer modeling of visual recognition, others are dissecting rat brains, absorbed in their studies of the neurochemistry of cortical neurons. Within every field, there is a snug and comfortable niche that matches a particular student's talents and passions; outside of one's niche—and this is true for everyone—the world can seem abstruse, overwhelming, or dull. Students feel very grateful that there is little necessity to stray far from "home."

♦ I felt very competent in my own tiny area, but I considered many of the other areas in my field to be beyond me. I felt very insecure because of this, and thought I would eventually be found out as a dim-wit! Then I attended a seminar on one of these difficult areas with my advisor, and he admitted that he didn't understand a word of it. I realized then that even professors aren't good at all facets of their fields—and they can still get tenure! (Ph.D., Psychology, biopsychology research emphasis, University of California, San Diego)

Academia is a world of specialists. While one can and should manage to stay abreast of major developments in outside areas, and should attempt to be a broad and synthesizing thinker, each scientist is still an expert only on his or her own version of the "crocodile's eyelid."

8

Networking

Maneuvering within the modern scientific arena takes much more than mere competence. In this respect, the world of science is no different from the world of business, politics, or other professions: *much depends on who you know and who knows you.* You may think that just doing good science is enough, that your publication record will speak for itself. Unfortunately, doing good work—although this is obviously critical—is not sufficient to guarantee future employment or posts of power and honor, such as that of journal editor or committee member. There are many more capable people out there than there are positions available.

It is only natural that when a position is open and a choice must be made, that the person chosen to fill the empty spot is someone that the decision makers know personally and judge to be capable. After all, why pick an unknown (except on paper) entity when an able friend or acquaintance can be chosen instead? It is for this reason that you cannot seclude yourself from the rest of the scientific community. People must get to know you, and you must get to know them. You cannot carry on alone and expect to advance too far—you have to network.

Opportunities at Your University

Your Advisor

The first important networking move that a student performs involves the careful choosing of an advisor. It is usually preferable to work for the best person in the field that you can, as coming from the lab of a distinguished scientist will put you in a select and respected club, which will ultimately be exceedingly useful to your career. The prestige of working for a well-known individual is so high, in fact, that a certain amount of privilege and respect is automatically awarded (deservingly or undeservingly so) to all members of his or her educational lineage, including the students of the individual, the students' students, etc. Being a member of such a "scientific family" often can make a significant difference in getting a job. However, most students do not have the option of working with celebrated researchers, and they have to push a bit more to be noticed.

Your Department and Outside Departments

Making contacts starts in your own department. Get introduced to the various faculty members as soon as possible so they know who you are. These people will be formally and informally evaluating you all the years that you are there and, yes, professors do gossip among themselves. Faculty members will be recommending you for financial support, postdoctoral positions, and opportunities later on in your career, so their perception of you is critical. Let the faculty and other members of the department get to know you. Join committees. Participate in departmental social events. Contribute to discussions. Collaborate on a project with faculty members inside or outside of the department. Organize seminars. Volunteer to be a student member on faculty committees so that you and the professors can interact as colleagues. Speak to fellow grad students in other labs. Do not hesitate to assist your advisor and other professors if they need your help: house-sit, check over grant proposals, pick up guest speakers at the airport.

No matter what the situation, *never* brag, put others down to make yourself look good, or rudely or patently push to get noticed; people will only think poorly of you. In a classroom, if you have an insightful comment to make, do so, but do not speak just to be heard. A choice student is friendly, involved, dependable, enthusiastic, dedicated, hard-working, careful, focused, optimistic, mature, confident, and modest at the same time. She handles difficulties well, rolls with the punches, and is noncomplaining. She also treats other students in a friendly, noncompetitive, and helpful manner. Is this you?

♦ I think that grad students need to form as many collaborations and begin networking as soon as possible. Work with professors in the same department or other departments, or even other campuses. You need to prove that you are a *team player* and that you can successfully work with a number of different people. You will build up your reputation, and get strong letters of recommendation when the time comes.

I'm happiest when I'm very busy so I joined the student chapter of one of the scientific societies. That proved to be a real smart decision. I was able to meet students and profs from other departments. It was fun and I learned how to manage people and organize events. (Ph.D., Chemistry, University of Rochester)

Self-promotion is a part of networking and, although it goes against the grain for many, it is a necessary part of surviving in science. To succeed now and later, you have to sell yourself to others. As you mature in your scientific abilities, you must become a strong and convincing self-promoter of your work and your thoughts. Be hardy and assertive, and do not crumble when challenged or provoked. Know why your work is important, and look for opportunities to describe your work to others. Draw attention to your strengths on applications or at interviews. Science is no place for the shy; you must hustle and be a go-getter. Be bold and determined and network with confidence.

Opportunities within the Larger Scientific Community

Students are directly exposed to a myriad of researchers in their field at seminars and professional meetings and, therefore, they should make every effort to attend these events.

Seminars

Most universities periodically invite accomplished researchers to their campuses to give talks on their work. This practice provides an excellent opportunity for students to listen to and meet with leaders in various fields. We encourage those students who are very interested in the work of invited guests, to read up on the work beforehand, and to try to talk to the scientists when their seminars are over. Introduce yourself, and convey your enthusiasm by complimenting their research. Even famous scientists are pleased when someone shows a genuine and sincere interest in their work. Ask intelligent questions (not ones whose answers can easily be found in the literature); such questions are usually specific in nature, although general inquires that show insight are fine, too. Feel free to discuss your own research if it is related to the speaker's work. Do not be pushy or clingy; other people want to speak to these interesting guests, too. If you are at the stage where you are looking for a postdoctoral sponsor, you may want to ask your advisor if a private discussion can be arranged.

Meetings

Professional societies in every field have annual meetings that attract members from across the nation or even the world. Your department or your advisor may be willing to pay your way to these meetings, especially if you are an advanced graduate student and can give a talk or present a poster. Try to attend as many meetings as you possibly can. You will find that they are great places in which to learn from others, experience camaraderie, and network. The atmosphere is congenial, people are in the mood for chitchat, and friendships are easily made. It is easy to start conversations at a poster, after a talk, or while waiting in line at the food service. People are always eager to talk about their work and are appreciative when others find it interesting, so do not be shy about asking questions or making comments. It is especially easy (and fun!) to share ideas and develop acquaintances with other students this way. You may want to get together with your new friends for lunch to continue your conversations, or you may invite them to visit your poster demonstration the next day. Be sure to exchange e-mail addresses with them, as this, too, is networking. Within a decade, these young people will be the new crop of assistant professors and industrial scientists. Friendships originally made during student days can be particularly strong and enduring and, besides providing pleasure and intellectual stimulation, can lead to collaborations, recommendations, and other forms of assistance.

♦   The first time I presented a poster at a national meeting was scary. For one thing, I was amazed at how long it took to prepare the poster—getting just the right photographs and trimming the prints so the important areas stood out (and dirt/artifact were hidden!), graphing the data just so, wording everything so it read well, arranging and pasting everything on the posterboard carefully so that it looked good—it took weeks and I thought I wouldn't finish in time (of course, *everyone* finishes their poster *just* in time)!

Then I worried about the meeting itself—would it be like an *oral exam?* Actually, once I pinned my poster up and stood there waiting to be attacked, I found that people strolled over and were very friendly and enthusiastic. They made suggestions about new experiments, told me about some references, wanted to collaborate, etc. It was easy to talk to them about my work. No one made me uncomfortable. (Ph.D., Anatomy and Cell Biology, University of California, Los Angeles)

Do not hesitate to also approach the more distinguished scientists at the meeting if you have something to say. Although getting access to these popular people may be more difficult, most of them are happy to interact with interesting and enthusiastic students, at least for a while. If a group of people surrounds the person you wish to speak to, stay close and listen to what they are chatting about. If the conversation is not a private one, hang around. Eventually you should be able to join in the conversation. If your advisor is playing his part, he will make a point of introducing you to some of the distinguished researchers in your area. If you are lucky, you may even be invited to join him and his friends for dinner. Dinner meetings are usually loose and happy, group events where a number of scientists get together to renew old acquaintances, enjoy themselves, and perhaps talk shop. They are excellent chances for your advisor to introduce you to, and bring you into, the "inner circle."

Between sessions, do not wait in a corner by yourself—be friendly and strike up conversations. In the evenings, do not take in a movie or order a pizza in your room—go to some of the scheduled socials. Joining one of the society's committees is a good way to meet people (and will even teach you a thing or two about politics in science). Remember that at a large meeting, eminent scientists will be all around you—you will be amazed at how famous that woman on the shuttle bus with whom you chatted about the weather, or that man with whom you offered to share a cab back to the airport, turns out to be! Why not have a cup of coffee with him while waiting for the plane?

♦ You rely on your advisor to introduce you to people in the field. I probably wouldn't have met any of them if my advisor hadn't helped me "break the ice." When I was a first-year grad student and at a national meeting, she said, "Hey, come to lunch with us." There were 2–3 people there who were doing projects similar to mine, and she said, "Why don't you tell them what you are doing?" And we all discussed it. (Graduate student, Neurobiology and Anatomy, University of Rochester)

◆ Networking begins at conferences. Don't sit in the hotel room, go sightseeing, or get drunk when you go to a conference—these are perks that established scientists enjoy. As a grad student you need to go to the talks and especially the poster sessions. You need to gather up your courage and go introduce yourself to as many people as possible. Express interest in their work, ask questions, ask for preprints of articles, get e-mail addresses. This is how to start a network. THEN when you need a postdoctoral position you can say, "Hey do you remember me? I met you at such and such a meeting and we talked about whatever. . . ." It works every time. People love to feel needed so if you have questions after the conference then give the people a call!! You will find that most people are happy to help. Networking and collaborating are the most important things to accomplish as a grad student. (If your advisor has no money to send you to conferences then start e-mailing the authors of the journal articles that you read.) (Ph.D., Chemistry, University of Rochester)

Increasing Your Contacts

There are indirect ways, too, that you can interact with the scientific community. If you read an article that you find particularly interesting, you may want to write a letter telling the authors how you feel. If your work relates to that of others, feel free to send them unsolicited reprints. If you have questions that you want answered, give people who can help you a call. Be sincere in your compliments and requests—do not contact people without a bona fide reason to do so.

The "In" Crowd

Although there are many scientists in the world today, the number of them working in any particular subfield of research is not very large, and the number of distinguished researchers in a subfield is smaller still. These are the people who are awarded large grants year after year; who have their pick of the best grad students and postdocs; who edit the journals and review submitted manuscripts (deciding who gets published and who does not); who serve on grant committees and determine which grant proposals are funded (deciding who is allowed to practice science and who is not); and who run the conferences and decide which scientists will present their work (determining who is thrust into the limelight and who is not). Thus, this group of scientists wields much of the power in the field. It is in your best interest to meet these people. You want to gain the name recognition that you need to prevent being lost in the scientific crowd. Also, be on the lookout for those who might make good postdoctoral sponsors. Associating with this community will increase the probability that those with influence will recommend you for the job—whether it be a faculty slot, a journal editor position, or a speaking engagement—when the time comes. Yes, the "old boys/girls network" is alive and well in science, too.

A Career-Long Activity

Networking starts early on at your own university as you win the friendship and respect of your fellow graduate students and your professors, and then branches out over the years to include people from across the country and around the world. Every contact is important. You can hope that these friends will be there for you (as you will be there for them) when you need information, advice, collaborations, recommendations, and invitations. Networking is a career-long activity. Do not get lost in the crowd.

9

Picking a Dissertation Committee, and Defending the Proposal at the Preliminary Oral Exam

There are two major oral exams associated with the dissertation research in most departments. For each, the student stands before the "dissertation committee" and "defends" the proposed, or the completed, research. The "final oral exam" (the "defense") is taken after the student has submitted his or her dissertation, and marks the end of graduate school and the awarding of the Ph.D. The other oral exam, known as the "preliminary oral exam" (some departments consider this exam part of the "qualifying" exam), is taken at an earlier stage, and marks the point at which the student is formally approved to embark on the proposed dissertation research. Some departments require only one oral exam, either the preliminary or the final. (See Chapter 22 for more information on the final exam.)

While passing the preliminary oral marks the committee's acceptance of the student's research plans, the student has actually been doing quite a bit—often a year or two—of research before introducing at this oral the exact experiments he "proposes" to conduct in depth for the dissertation research. These preliminary studies serve to find a line of experimentation that seems promising and technically doable; usually some positive findings have been noted that make further investigation of the issues, in terms of dissertation projects, encouraging. Projects that were started, but that proved disappointing in terms of feasibility or promise, are left by the wayside and never make it into the dissertation pro-posal; sometimes such studies are actually complete enough to re-sult in a paper, before being dropped for being dissertation "dead ends." It is the student's advisor who decides whether the early, pilot experiments are auspicious dissertation material. With little research experience, students may find it difficult to judge the promise or potential difficulties of a specific study.

The Proposal

After consulting with the advisor and deciding upon a viable series of proposed experiments for the dissertation (many of which have already been started in bits and pieces, as stated above), the stu-dent begins to prepare for the preliminary oral exam. It is neces-sary to first write the proposal, a detailed description of the exper-iments that the student proposes to perform. Included in the proposal, which is usually similar to a grant application, are a re-view of the background literature with bibliography, explanations of the relevancy and meaning of the proposed work, an account of preliminary results, and a detailed description of the experimental designs and methods to be used. This is also the time to consider the statistics that will be needed, a factor that will influence ex-perimental design and numbers of animals used (consult a statisti-cian if necessary). The student should consult often with his advi-sor during the proposal writing. The size of a proposal varies, but five to twenty-five pages is common. It would be helpful to con-sult copies of old proposals to see the format employed (see Refer-ences for books useful for creating a proposal).

While a proposal requires a large amount of preparation and work, it is one of the most important pieces of writing that the student performs during his graduate years. There are three rea-sons for this. First, it forces the student to think deeply, carefully, and methodically. In fact, it may be the first time that the student has had to go through the rigors of designing complicated experi-mental protocols. Going through this process is an invaluable les-son in scientific reasoning. It also clarifies the experiments, eluci-dates any preliminary studies that may be needed, and brings to light possible conceptual/technical problems and errors, all of which go to show that you never really *know* an experiment until you try to write it down precisely. Writing the proposal may also bring to mind additional experiments or new approaches. Second, working out the details of the methods section will result in the production of what is essentially a "lab manual" that can then be followed, step by step, during the actual performance of the experiments. Third, the proposal can form the framework of the dissertation down the road: the background literature, methods, preliminary studies, and other sections can easily be transferred to appropriate chapters in the dissertation. Thus, much of the proposal can be reused or recycled, an observation that should make its writing less of a chore.

The Preliminary Oral Exam

After the advisor approves the final version of the proposal, the student gives a copy of it to each member of his dissertation committee to read before the preliminary oral exam. It is up to the student to arrange when the exam will be held, a duty that requires finding a block of time and a day that each member of the committee has free; as the exam room (usually some conference room) also must be free at this time, the coordinating task is tricky, and should be done long in advance. On the big day, the student is first asked to step outside the room while the committee reviews the student's progress and academic record. The student is then brought inside, and standing in front of the committee, gives a talk, which can vary from fifteen minutes to an hour in length, summarizing the main points of the proposal. Included are the background of the problem, the questions the research hopes to answer, the reasons that these answers are important, and how the research will be conducted. The student typically makes use of slides, or transparencies on an overhead projector, to illuminate the talk. During or after the presentation, the committee asks pertinent questions of the student. When the queries, which add at least another hour to the exam, are exhausted, the committee decides if the student has passed (i.e., if the student can proceed with the proposed research).

The preliminary oral exam serves a number of functions:

1. It forces the student to finally focus, months or even years after testing the waters and experimenting in a number of different areas, on serious projects that should lead to the granting of the Ph.D. By preparing the proposal, the student clarifies the projects in his mind, and works them out in detail.

2. It enables the committee to judge whether the proposed projects address important scientific problems, and whether the student sufficiently understands the studies about to be undertaken. Has the student thoroughly considered the scope and depth of the studies? Does he really know what is involved, technically and methodologically? Is the background literature in the field understood? Are the concepts and technical requirements commensurate with the student's understanding or abilities? Will he be able to interpret and explain the results procured? Does the student grasp the meaning of the work? If the answer to any of these is "no," the student may not pass the exam.

3. It enables the committee to make changes in the design or methodology of the studies, if need be. Are the conceptual framework, design, methods and analyses integrated and suitable for the specific aims? While the necessity for major revisions will likely result in failure of the student to pass the exam, insistence on minor changes helps ensure that the experiments are solid. If the student picked his committee wisely, there should be enough expertise spread among the group to assure the adequate evaluation of the various experiments proposed.

4. It safeguards the student. Passing the preliminary oral is like signing a "contract" with the committee: if the studies are dutifully carried out as agreed upon, the committee, for its part, should not accost the student with new research demands. Unfortunately, because research results are unpredictable, and research very often goes off in new directions unanticipated at the time of writing the proposal, the stipulations of the "contract" sometimes become a gray area. This is one reason why the student should continue to consult with the members of the committee several times a year.

Picking a Committee

Considering the seriousness of the preliminary and final oral exams, it should be obvious that picking the professors who will serve on your committee is a weighty decision, somewhat akin to picking an advisor. Bad advisors and difficult committee members can make a student's life miserable. Good advisors and helpful committee members can promote a career. Here, then, are some things to consider when selecting your committee:

1. By the time you get your degree, your committee members will know you quite well; they will know how you think, how you work, and how you deal with the difficulties of research. They are, therefore, obvious choices to list as references or to ask for letters of recommendation when the time comes to apply for a job. Some students think it is to their advantage to choose at least some committee members who are well known, respected, and politically connected as their recommendations should carry much weight. This is a personal choice, and something to think about.

2. Sometimes, a committee member demands modifications to a student's research plans that are unfair or unreasonable, and that could add months or years to the work schedule. In such a situation, it is hoped that other committee members will fight for the student, and overrule the difficult person. This has the potential to turn into a power struggle. Theoretically, the chairperson of your committee should be on your side, although even a chairperson can be intimidated by stalwart committee members. It is therefore advisable to have on your committee professors who already view you in a friendly light and who are known to be reasonable.

3. Go about picking a committee similarly to the way that you chose an advisor: speak to other students. They can tell you of their own experiences with committee members. Find out which professors have reputations for being fair and helpful. Avoid those who are out to prove their power over others, or those who may want to show off their intelligence and put you down in the process. You may be able to get advice from your advisor, but be sure to confirm any recommendations with students. If you choose a committee member who proves unusually difficult, it is possible to replace him or her with another professor even after the preliminary oral exam.

♦   I was stressed for my oral but I figured this was my opportunity to have the undivided attention of my committee and to explain to them what I was doing, how and why. Very few people flunked their oral exam. Even if someone failed they were usually given another chance and then allowed to stay in the program. My committee was very fair and I knew that before the exam started (from speaking to grad students that had the same profs on their committee), so I wasn't intimidated or worried about being humiliated or attacked. (Ph.D., Chemistry, University of Rochester)

4. It should be obvious that only those professors sophisticated in the general, if not specific, research areas that you are studying are qualified to judge your work. There is another reason, however, to include on your committee members with expertise in your sphere of experimentation: you may run into theoretical or technical problems as your work progresses, and you can turn to these professors for advice. This is especially important if your research delves into areas beyond the expertise of your advisor. Knowledgeable committee members are necessary to prevent you from making conceptual or experimental errors, and checking in with them a few times per year protects you from receiving any unpleasant surprises at the final oral, such as hearing that you performed faulty research. (Some universities mandate one to two formal meetings of the student and the committee each year. These leave a paper trail in the Dean's office that can aid the student should questions about the research arise in the future.)

Universities differ in the rules of committee composition. In most cases, your advisor automatically becomes the chairperson of the committee. The chairperson handles committee quarrels, and usually stands up for you if someone on the committee makes unreasonable demands; sometimes, the reverse happens, and the rest of the committee has to protect you from your advisor (for instance, if they think he or she is keeping you on too long). Most universities require that at least one committee member have an affiliation outside of the student's department; some require that a member be from another university.

Anticipating Questions on the Exam

No matter how well you think you know your proposed experiments, it is important that you sit down with pad and pencil and go over the proposal line by line searching out statements that may spark questions in your committee member's minds. What is the evidence that the assays you plan to run are *specific* for what you are assaying? If you remove gland X, how will you know that you got it all out? (Are you planning to do any tests to show that you did?) Is the test that you plan to run really sufficient by itself to enable you to draw a conclusion? If you get result X (or Z or Y), how would you interpret that, what would it mean, and what experiments would it make sense to do next? What result do you *think* you will get? (What evidence is there in the literature that makes you think so?) What made you decide upon the particular statistics that you plan to use? Are you sure that everything that you say you are going to do is technically or practically feasible? What problems might you anticipate? If you indeed run into problems, what alternative tactics or experiments will you turn to? Make sure you can back up any statements that you make in the proposal. Ask your advisor if there are weaknesses in the proposal that you may be targeted for, and jointly go over answers that will address questions on these issues. A professor writes:

> ♦ Yes, the exam is difficult, yet students have *more anxiety than they should*. The exam is used to evaluate thinking processes. Students are anxious because some do fail every couple of years. (Ph.D., Biophysics, University of Rochester)

Make sure that you also contemplate in advance your answers to those "general" types of questions that committees love to ask: What will this work add to the field and why is it important? What do you hope to learn? What made you interested in the research? (They sometimes throw in personal questions, so do not get flustered.) Know the specialties of your committee members, and anticipate that they will ask you questions on topics or techniques mentioned in your proposal about which they are particularly knowledgeable. Find advanced students who will share with you the questions they were asked at their preliminary oral. Practice your talk in front of students, or your lab, and have them ask you questions that a committee member might come up with. Being able to answer questions from this mock committee should give you confidence for the real thing. A little caveat: although the exam questions are almost always in reference to the proposal at hand, officially, *any scientific question can be asked at a prelimi-*

nary oral (some departments have a preliminary exam that covers the proposal plus a specified outside topic).

◆   A neuroscience student told me that at his preliminary oral a committee member asked him about the *greenhouse effect!* How can anyone prepare for random questions like that? I decided to forget about trying, and only concentrated on my own research area. I was lucky—I *was* only asked about my research at the exam. (Ph.D., Psychology, biopsychology research emphasis, University of California, Los Angeles)

Some questions that are put to you will end up being answered or debated by the committee alone, without your input at all! After all, the committee is there to make sure that your projects are going to be sound, and sometimes they need each other's help in this. If you are asked a question and have no idea of the answer, never try to fake it. Admit that you do not know. However, if you can offer any relevant information related to the question, mention that you do not know the specific answer, say what you do know, gently direct the discussion onto the next issue or ask if there are any other questions, and hope that you can get away with it. (Responding to a question by a half-answer that only peripherally relates to the question is a strategy known to most scientists on the lecture circuit!) Not all questions will be to test your knowledge; some will try to make you think, and hopefully will prod you to grasp certain nuances of your research. Think of the preliminary oral not as an exam, but as a meeting of colleagues who are trying to help you with your research; this attitude is usually an appropriate one. Try to keep your cool throughout the exam, and don't get intimidated; if you truly believe the committee is wrong about something, politely stand up for yourself. (Read the section in Chapter 18, The Final Oral Exam, on being in control of the exam situation—what applies to the final exam applies to the preliminary, too.)

◆   My preliminary oral involved both my own research, which was a breeze, and an outside topic. The paper I chose for the outside topic was very interesting, but as I read it more carefully I found that practically every equation in it was wrong, which led to some hilarity at the exam. There was one examiner whose questions I could never understand, but I learned later that hardly anyone else on the committee could either. I do recall one wonderful moment when one of the examiners asked me a series of questions whose point I couldn't quite get, and then, suddenly, I saw where he had been leading me and had one of those shining "Aha!" moments as I put it all together and reached the conclusion he wanted me to see. In all, I felt I did a good job on the oral, but felt a bit cheated that I hadn't had the opportunity to show very much of what I knew. (Ph.D., Chemistry, University of California, Berkeley)

◆   We had a preliminary oral exam, but did not have to write a proposal. The exam was not stressful. I had already started working on the project, and I said basically "This is what we are looking at, this is what we want to look at." I prepared hard for the presentation; I knew my stuff. I remember that one of the committee members said, "Well, why don't you start and tell us a little." They all asked a bunch of questions, and they were satisfied that I knew what I was doing. It seemed that if you took it seriously, if you walked in prepared, if your project was OK, there wasn't anything to be worried about. They weren't trying to be hard on you; they kept it pretty light. Obviously, you don't know as much as when you take the final exam, and if they wanted to be hard on you, they could have. (Ph.D., Theoretical Physics, Brown University)

It may be customary for you to bring in snacks for your committee to consume during the exam. It also is important for you to dress appropriately for the occasion; if tradition dictates, "dress up." Of course, you may be the only well-dressed one in the room; your committee will likely be wearing the typical casual clothing that one can get away with in academia.

10

The Life of a Graduate Student

The First Few Years

What is the world of the budding scientist like? For students in their first few years of training who are not working in the lab or field, life is not totally different from that of an undergraduate. Students attend classes, spend hours in the library gathering articles for papers that they have to write, work on problem sets, study for exams. They are challenged to think more deeply, carefully, and creatively than they were when they were undergraduates, however, and memorization is not emphasized. They likely spend a few hours each week attending seminars and journal clubs. Characteristically, they take knowledge very seriously and are searching for questions that they want to devote themselves to in later years.

Students who are also expected to participate in research during these years may start out by working on simple projects or by helping more advanced researchers; temporarily assisting others is a good way to learn techniques and scientific thinking, and students who work long and hard at this may have their efforts rewarded by middle authorship on the published studies. Students do not really have to know much about a field to physically start researching at a "cookbook" level. Many are actually unaware of the meaning of their work until they later fill in the gaps in their knowledge. One of the ways they catch up is by reviewing the relevant literature (available online or at the library). Deciding which articles to read for research purposes will come more or less naturally. In order to understand what is going on in the lab, students will want to read papers that have been published by current and former members of the lab, and those on related matters that have been published by other labs; they will also need articles describing the techniques they are thinking of using so that they can perform or modify them. By keeping up with the latest developments (reading the most recent articles), students can judge what experiments make sense to run in the future.

When a student is not in class or in the lab or field, he is likely to be in one of three places: at a talk, in the library, or at his desk. Every graduate student is assigned a desk and often a file cabinet in either an office or the lab. This becomes his "home base" throughout the years of graduate school, a place to read journal articles, study, design experiments, compose manuscripts, relax, search through *Current Contents* (a listing of the contents of current journals) etc. For theoretical scientists, the actual research may be done at the desk, too. (If a student does not have a computer on the desk, some of these activities will be carried out wherever the lab's computers happen to be located.)

♦   My desk is the area of the lab that is mine—the area that no one touches. My working desk is really not there—it's in front of the computer. So my desk serves as the place where I store everything, where I have everything organized the way I want. My desk is where I can make piles—as many piles as I want—so long as they don't fall off the edge. (Graduate student, Chemistry, Vanderbilt University)

Students who engage in research during the coursework years must do a lot of juggling to handle all their responsibilities successfully. Although they will be busy with classes and related activities, they are still expected to spend a considerable amount of time in the laboratory (or in the field); it is almost as if coursework were considered something they take care of "on their own time." Still, it is understood that there will be days when they will be too busy with coursework to come to the lab, and no explanation for their absence will be necessary. As long as their research is progressing, advisors are not going to mind. The system should not be taken advantage of, however; research is a top priority, and as much time and effort as possible should be devoted to it—it is a graduate student's raison d'être.

Teaching Assistantships

Students may be supporting themselves during these or later years by working as teaching assistants. Many departments, in fact, *require* their graduate students to be TAs for at least one semester so that they gain teaching experience. At the same time that these students are struggling with classes and perhaps research, they are also grading exams, preparing lectures, holding office hours, assisting in a lab course, etc. At the same time that they are worrying about their own lives as students, they are assisting others with theirs. Nevertheless, many students *love* TAing. Teaching is fun, and being an authority figure is ego-boosting. However, TAing can take a lot of time away from other things so, although it looks good on a résumé and carries some weight when one is being considered for a faculty position, *you do not want to do it more than a few times, if possible.* Good science teachers are generally underappreciated at a research university (this is true for professors and TAs alike), so do not expect to labor over your teaching responsibilities to the detriment of your research and have your efforts appreciated.

♦ Teaching was fun for the first semester or two, but it became a real time-consumer later. I would say, avoid it if possible. (Graduate student, Neuroscience and Behavior Program, University of Massachusetts, Amherst)

The Later Years

There is a certain predictability to the lives of advanced graduate students in the laboratory sciences (and even more so for those in theoretical specialties). The days of a student who has finished coursework are devoted basically to the daily matters of research; peripheral to this are visits to the library, periodic journal clubs and lab meetings, and the occasional seminar. This is the backdrop against which the student squeezes in all other appropriate activities: studying for the qualifying exam, analyzing collected data, researching and writing the proposal, preparing for presentations at professional meetings, composing journal articles, meeting with committee members, and writing the dissertation.

♦   I would get in around 7:30 a.m. Sometimes I would sit down at the computer and do modeling. Or, I might work on writing a journal article. If the computer had been working on my simulations from the night before, I'd see how they ran. If I had been programming a lot, and the programs weren't working, then I would go and talk to my advisor; sometimes the error wasn't a mistake in programming, but an error in our concept of what the model should be. These activities were pretty typical of my days. *All* of my work was on the computer. (Ph.D., Theoretical Physics, Washington University)

♦   On the day of an experiment, I take mice from the animal room and bring them to the lab. The day before I labeled hundreds of test tubes, and now I get out the glassware, instruments, ethanol, and other needed materials. Each mouse is sacrificed, and the necessary organs are removed and ground into cell suspensions. The cells are counted, divided into groups of equal numbers, and treated with different concentrations of various reagents. This is precise and tedious work, and sterile technique is used. At the end of the day, I am glad that all the tubes are gone and everything is put away. I am usually wiped out. (Graduate student, Biological Sciences, Columbia University)

♦   A typical working day for me is 8–10 hours long; occasional 12-hour days are not that unusual, nor is spending 4–6 hours in the lab on a weekend day.

I have spent many months at a time where I did nothing but sample preparation. This involved a lot of 12–15-hour days in the lab. On those days my entire schedule—when I came in, when I left, even when I went to the bathroom—was dictated by the experiments (HPLC chromatography) I was doing. That was a very stressful time. The work was dull and tedious. Experiments failed as often (or more often) than they worked, and the hours were very long and tiring.

I have also spent a lot of time doing nuclear magnetic resonance (NMR) experiments. During those periods my schedule was very busy, but not nearly as difficult and frustrating. The equipment dictated my schedule to some extent, but in fact once the conditions were properly set up, the computer pretty much took control, and I was free to focus on other things for the next 36 hours or so. NMR is a technique that I actually enjoy, so I did not mind the somewhat unusual hours.

Now that I am finishing up, I am doing only data analysis and writing up my results. My days are spent mostly at a computer rather than a lab bench. I am finding this to be more boring than I had expected. Furthermore, I have no lab experiments or equipment to dictate when I must work, so it is challenging to keep up my motivation and get the work done. (Graduate student, Biophysics, University of Rochester)

Each researcher in a lab is probably working on a separate project, or a separate part of a project, and each organizes her own day, and runs around doing her own thing. At any moment, one person may be on the computer, another is tinkering with equipment, still another is plating cells, one is in the cafeteria, one is studying for an exam. The atmosphere is generally casual and informal. There are likely to be moments of hearty laughter. People alternate between being sociable and talkative when not too preoccupied, and being totally absorbed and uncommunicative when concentrating. Jeans and T-shirts are typical attire—it is a comfortable environment to work in!

The lives of students engaged in field work are considerably less predictable than those of their laboratory-oriented peers. While they, too, participate in departmental and lab activities when they are on campus, their days in the field are highlighted by adventure, danger, hardship, physical exertion, and particular problems not encountered by those who work behind walls. They patiently endure the long stretches in which nothing is happening, the discomfort of being at the mercy of the elements, and the frustration of not being able to carry out experiments planned earlier back at the university.

♦ I went to graduate school for adventure and discovery. I made sure that I was able to combine my scientific interests with a lifestyle I enjoyed. I remember most the time that I spent in the field looking for samples of magma for geothermal analysis. On one such trip, my job was to acquire samples from a part of the Aleutian Islands that was very inaccessible; there were not even footpaths or bridges to aid us. I had to arrange transportation to remote sites near our destination (we were often aided by U.S. military transportation and mail vehicles). After several bus rides and military flights, my guide and I were dropped off by helicopter in an open field, and we knew only that we had six weeks to get to our destination, find our samples, and return to a prearranged pick-up site. The clock began ticking as soon as the helicopter took off. I realize, in hindsight, that such restrictions prompted us to compromise our safety. I recall going out in severe storms, after spending a week twiddling thumbs, because we could not afford to wait in our tent for the weather to clear. Similarly, I recall wading through streams while loaded with gear, crossing at convenient rather than safe spots in order to save time. On one occasion, while shoulder deep in ice-cold water, my guide slipped and fell; fortunately, I was able to rescue him and recover most of our gear downstream.

The sense of accomplishment that one feels, overcoming the obstacles of nature such as fjords, streams, and mountains, provides an intense thrill that is rivaled by little else. Battling the climate also provided experiences that I continue to view with great pride and satisfaction even today. Overcoming nature's obstacles often provided a better "rush" than harvesting the samples that were the scientific aim of the project. I think that the thrill of adventure in the remote wilderness brought me as close to my primordial self as one can get in today's society. Man against nature—WOW! (Ph.D., Geology, University of California, San Diego/ Scripps Research Institute)

♦   The best way to learn to do field work is to work first with somebody who is experienced in the field. Some people are not very good in the field and won't ever be, others are natural at it—meaning that they get a lot of data in a very efficient way. It is best to learn by watching and participating in someone's field project. Failing that, take a course or two at a biological field station, since almost all stations offer courses that involve field work and many also have senior investigators doing original research that one can observe. (Field stations vary tremendously in the rules and guidelines for doing research—where research can be done, how many animals/plants can be sampled in a particular area, etc. Contact each field station separately for their specific set of rules. In many cases, a research outline must be approved by the station before admission is granted.)

Hopefully, the person in charge of any field work has made it clear what is expected in terms of number of workdays per week, conditions in the field, who pays for what, who does the cooking and dish-washing, etc. Students with a tendency to grumble are hopefully weeded out before being taken on.

In some parts of the country or world, it is inadvisable to work alone. Sometimes working alone is unavoidable, but safety precautions should be taken (carry cellular phone, alarms, etc.). Sometimes the danger is not from other people but animals (grizzlies, snakes, etc.) or hazards (falling in a pond while wearing waders and not being able to get out). Be sure somebody knows where you are and when you should return. Learn from others who have worked in similar circumstances what the likely hazards are and prepare ahead of time for an emergency.

The major difference between field work and lab work is that data generation is usually considerably slower in the field. You spend much more time traveling, maintaining yourself (arranging for meals), hiking to field sites, and other time-consuming activities that don't generate data. Furthermore, field work is very seasonal, whereas most lab work can be done year-round. Therefore, a field-based project may take 2–4 times the amount of time to generate the same amount of data as a lab project. There isn't any reason why the quality of work can't be as high in the field as in the lab. The scientific method is still applicable.

Hardships are frequent. Bugs bite, the sun burns, the cold wind blows, vehicles break down, equipment breaks down, terrorists attack, etc. One of the most important attributes of a good field researcher is that she is able to put up with alot of discomfort without complaint.

BUT—field work keeps me alive. Being out in the field with the animals you are working on is essential to the design of a good lab study. If you don't experience what they do, how can you understand their physiology, behavior, or other attributes? Keep your eyes open. In contrast to the lab, you may see interactions in the field that open up new avenues of exploration and raise new questions. Avoid making rules, like "this type of sparrow never builds its nest in that type of habitat so I won't look there for nests." Every rule I made was broken by the animal I was studying. Keep an open mind and learn. (Ph.D., Zoology, University of Michigan)

The Nature of Research

♦ I love learning how things work, how nature is put together. This is what attracted me to science. I wanted to learn how the body works. (Graduate student, Zoology, Duke University)

Graduate students have been granted a special favor. They have been allowed to pursue their fantasies. They are given the freedom to explore how nature operates, and they do so through the eyes and ears and minds of the most sophisticated instruments and technologies mankind has to offer. Commanding these powerful tools, they ask interesting questions and get interesting answers. Money is provided for them, so that supplies can be bought and experiments conducted. Not burdened with having to write their own grants, and given access to marvelous facilities, they are free to be full-time researchers, discoverers of new things. Theirs is truly a privileged world.

Graduate students know how fortunate they are. Experimentation is an outlet for their natural curiosity, and they revel in being part of a community of scholars and seekers. Overall, a sense of excitement and possibility surrounds a research university, and it filters down to every person who is on the campus. Still, graduate school is not utopia, and many of the practical aspects of research are not particularly glamorous. Certainly, students will have glorious moments when something remarkable, or just plain noteworthy, is first observed, but such occurrences will make up only a fraction of their research hours. The remaining hours, to be realistic, are more likely to be filled with activities that become quite mundane over time: wearisome procedure after procedure, detailed observation after observation, precise manipulation after manipulation, complex analysis after analysis.

♦ I did not anticipate how detailed and tedious the day-to-day work of scientific research had to be. For every question, I had to show half a dozen appropriate controls, or else the answer was useless! The procedures I was using were multistep, sometimes taking days to complete. At every step, there was the possibility of an error which could force you to start over from the beginning.

There were many times when I was less than enthusiastic about my research project! Everyone gets frustrated with their research; that's just how it is. There were probably half a dozen times when I was sure I was going to quit. I would often say, OK, I'm just going to finish this part of it, and then I'm out of here. I would characterize graduate school as being 90 percent tedium and frustration and 10 percent discovery and jubilation. Somehow, that 10 percent has to carry you through the other 90 percent. (Ph.D., Biology, University of Texas, Austin)

♦ If I am doing immunocytochemistry, my day starts at about 9 a.m. I put the tissue in buffer for 10 min, then remove the buffer and add fresh buffer; this is repeated 9 times. I then incubate the tissues in NGS for 30 min, and in secondary antibody for 45 min. I pull the tissue out of the wells, and rinse them 9 times for 10 min each. While they are rinsing, I mix up the AB solution. The tissue is incubated in this for 1 hr, then rinsed 9 times for 10 min each. It is then put into another buffer for 20 min, while I make up the DAB solution. Then I react the tissue. If the procedure works, I look at the results. I go home at 5–6 p.m. I make dinner, eat, read/work/watch TV, and go to bed around 11. (Graduate student, Anatomy and Neurobiology, Washington University School of Medicine)

♦ We made grids on the ground and had to count all the various types of seeds that we found there. It was slow, painstaking and painfully boring work. The exciting discoveries that they give Nobel prizes for happen to only a minuscule percent of researchers. For me, every little success kept me going. (Ph.D., Biological Sciences, University of Michigan)

♦ Researchers work long hours to make small strides; the *small* rewards are highly memorable. One joyful occasion included the day that I actually began an experiment that I had spent months designing and researching. Successful troubleshooting brought frequent bursts of elation in the months that followed. (Graduate student, Neuroscience and Behavior, University of Massachusetts, Amherst)

While the boredom can be eased somewhat by alternating research with writing or other duties, the research grind can still be a source of great frustration and stress. Students may feel at some point that they have nothing to show for their work, or that their work is of no consequence. "Disasters" will occur: animals will die, plants will be washed away by rains, equipment will break down, chemicals will go bad, errors will be made, shipments will be delayed, etc. Some experiments may even have to be totally dropped because of problems.

♦ Because of technical errors in the equipment set-up (made prior to my entering the lab), I had to virtually redo my entire dissertation work. Data for a couple of papers also went down the drain. This was extremely frustrating and depressing, and there was little help/support. It killed a lot in me: self-esteem, confidence, optimism, motivation. (Ph.D., Neuroscience, university withheld upon request)

The frustration and/or the monotony of the research and the amount of time necessary to gather data varies from field to field. In some areas, particular studies are straightforward and quick, and it is possible to generate enough data for a paper in one day; in other areas, it can take years. Thus, some students receive slower feedback and/or have a more exasperating experience in graduate school than others. For many who have a difficult time, sheer determination is necessary to get them through.

♦   I did not get along well with my mentor, but I had a great deal of interest in my research which is the only thing that kept me motivated for five years. Try not to get too comfortable with the "job" and life of a student or you will never graduate. Many times I was discouraged and disenchanted with the whole thing but I knew I had to keep focused on the goal and not let the daily grind of things get me down. My mentor was not very motivating, and rarely had a good mood. He spent his waking hours at the computer rewriting grant proposals, which taught me that that was *not* what I wanted from life. I love teaching and am glad my Ph.D. took me to that career. (Ph.D., Biology, University of Pennsylvania)

The Saving Grace

Yes, research is tough. It can be monotonous. It is frustrating. But there is *nothing* like it. Nothing compares to the sheer joy of the discovery process. Thinking about the workings of nature and planning experiments to test ideas is exciting and stimulating. Operating state-of-the-art equipment and working with exotic and/or difficult-to-obtain natural material is a privilege. Preliminary results can be tantalizing. Finding something new is ecstasy. That initial moment when the data fall into place or the observation is made and you know you have something—something that no one else in the history of mankind has ever witnessed before—is pure enchantment. This is why people do science and put up with all the negatives. Science is a thrill like no other, a giddy treasure hunt, a euphoric rocket ride. Nothing else so serious can be so much fun.

♦   What gave me the most satisfaction was knowing that I was doing or seeing something, even if small and trivial, that NO ONE had ever done or seen before—for example, using an electron microscope and seeing certain structures in the cells I was studying which no one had ever seen in those cells (or at least not published) before. I felt a great sense of wonder at the complexity of the things I was studying—it reinforced my reason for doing science. (Ph.D., Biology, University of Texas, Austin)

♦   I know from experience that lab work can be monotonous and boring and that 95 percent of research gives no results. It is very frustrating because you have to work out techniques, etc. It takes two months from the time I start an experiment until I see results; for repeated failures, it can be a downer. It's the 5 percent of the time, when the experiment works and the results are fascinating—when my advisor and I will stand over them and say, "Wow, this is really cool"—that makes it all worthwhile. That's what keeps you going. (M.D./Ph.D. student, Neurobiology and Anatomy, University of Rochester School of Medicine)

As students become more involved in their work, dissertation matters are always in their thoughts. Ideas about research crop up everywhere, and during every activity.

◆   Research dominated my mind even when I wasn't in the lab. Spectra danced through my dreams, and sometimes I would wander around campus in a sort of daze trying to solve some paradox related to my research. At one time I could reproduce from memory dozens of spectra involving hundreds of vibrational frequencies with reasonable fidelity—as could my advisor! We could, and often did, hold lively discussions about the assignments of those dozens of spectra without a single piece of paper passing between us. (Ph.D., Chemistry, University of California, Berkeley)

The prospect of publishing one's findings in scientific journals is a major motivator for students. Enormous satisfaction is garnered from being published, as is a sense of completion, and a renewed vigor for further experimentation. All students are proud to see their names on articles in respected journals.

◆   I kept staring at my name there on the page. My name, associated with that esoteric-sounding title of the article. What a thrill! It was one of the great highs of grad school. (Ph.D., Geophysics, Stanford University)

Most people like to publish as soon as it is reasonable to do so, but overly compulsive researchers may hold off on publishing until they have replicated their results again and again and again, and have added every control that can be possibly imagined. There *is* such a thing as being *too* careful—you do not want to publish only a few papers during your career. It is not a total disaster if you unknowingly publish something that you (or others) later show to be incorrect; it happens all the time. This is not to say that you should be a sloppy researcher; be cautious and mindful, but realize that errors occur and that results and conclusions are not etched in stone.

How Much Work Is Enough?

One of the authors remembers going to a graduate school interview and being struck by the words of one of the professors: "I worked all the time when I was in graduate school at MIT; I never knew how much work was enough."

No one is likely to tell you how hard you are expected to work as a student. The silence on this issue is especially uncomfortable. Surely academia is not a nine-to-five job, and there will always be times when working through the night (and weekends/holidays) is necessary, but is nine-to-five OK on some days? And how long is a *typical* day?

If you stand outside the buildings of any major research institution in the middle of the night, you are sure to see a sprinkling of lighted windows. Labs can be used around the clock, and there are always some souls who toil away into the night and go on to see the sun rise. People in the sciences keep very unusual hours. You will find that some like to sleep until late in the morning, and then come to the lab to work until late into the evening; others may start their working day long before the sky is lit. Such irregular hours are usually tolerated well in academia, as long as productivity is the outcome (new students should try to keep more regular hours, and work when others are around). Sometimes the nature of the experiment demands an eccentric time schedule; many neurophysiologists, for example, work late trying to record as long as possible from the cells they have impaled:

♦ I worked long hours—each experiment lasted 14–20 hours, and we did one or two per week (plus follow-up work). I also worked as a TA for a lab section. (My boyfriend thought I was having an affair when I came home at 3 or 4 a.m.). My research clearly dominated my life, and I liked it! (Ph.D., Neuroscience, university withheld upon request)

♦ My advisor used to tell me, "The animals don't know the meaning of a weekend or a holiday." (Ph.D., department withheld upon request, Johns Hopkins University)

Other researchers who work late may simply be workaholics, who return eagerly to their labs after dinner and can get much accomplished in the relative quiet of the night hours. Some may be over-tired souls who would rather be sleeping, but are driven to get data for a meeting or a paper. Still others, like the professor quoted above, just do not know "how much work is enough" and are motivated by the uneasiness that that creates.

♦   One of the projects required our presence from 6 a.m. to 8 p.m., collecting hourly samples. Sometimes we had to come in during the weekends also. Some days, I used to sleep at the department—for one experiment I slept for three months in my office so that we could start an experiment early and complete it early. My colleague used to come in at 4 a.m. so that he could analyze samples. These time schedules occurred periodically over the five years, but were not a regular feature. We did not follow these timings under instructions from our advisor—we did it on our own. This life may or may not be typical for a graduate student. (Ph.D., Physiology, Kansas State University)

In truth, we cannot advise you how many working hours are appropriate. This will depend upon the lab you work in and the personality of the lab head, the type of work you do, and your drive. We can, however, make certain observations:

Every lab exhibits a different working environment. Some pressure their members to put in long, exhausting hours; others are more lenient, letting individuals set their own working pace. Sometimes the pressure to work excessively comes from the lab head, sometimes it comes from the excitement of working in a hot field where the competition is intense.

♦   I don't feel like I worked long hours . . . I occasionally worked weekends. I worked because I wanted to get things accomplished, not because it was expected or demanded. If I needed to use shared instrumentation and it was only available at certain times, then I would work nights or weekends, etc. Sometimes it was nice to come in on weekends because there were fewer distractions. (Ph.D., Chemistry, University of Rochester)

♦   My advisor worked me hard, but I got out a lot of papers, so it was probably worth it. I know of some advisors who work their students beyond the bounds of healthy toil—even when the students are publishing consistently. I think there is a point where too much is just "too much." (Ph.D., Psychology, biopsychology research emphasis, Purdue University)

Long hours are sometimes the result of poor organization or poor time management. If you plan your experiments appropriately, make efficient use of free blocks of time, and are well prepared each time you tackle a project, then you can accomplish the most in the minimum amount of time.

While grad students may work long and hard, they also have the luxury of being able to come and go when they please. If they have to leave at noon one day to go on a personal errand, no one is going to care. If they work all night on an experiment, they are free to take the next day off. If they accomplish little one week, they can make up for it the week after. *This absence of "clock-punching" is one of the great advantages of working in academia, whether one is a student or a professor.* Work is so much more tolerable when it is a matter of choice. It sounds as if it would be common for students to take advantage of a system with freedom like this, but few ever do. After all, students must work hard to be productive, and they know that productivity is the name of the game.

♦   You don't know how many hours you should be putting in each day. As long as my advisor is happy with me, and tells me so (or if she responds positively when I ask her how I am doing), then I assume that I am working hard enough. (Graduate student, Biochemistry, Rutgers University)

♦   How hard do you have to work? As my advisor told me, "You have total freedom to work any 90 hours per week." (Ph.D., Biochemistry, Division of Biology, California Institute of Technology)

♦   One thing that sticks in my mind is that I always felt like everybody else was working harder than I was. Not because I wasn't working hard, but because I thought I wasn't working hard enough. Then I realized that doing research had its own ebb and flow—for a week or so (or maybe more) I'd be puttering around not getting much accomplished, but then I'd spend a couple of weeks working like crazy. I think it's important to get comfortable with your own manner of working, and also to understand that "down times," when you are not actually working frantically at the bench, are very important, too. It's a time when thoughts and ideas come together on their own without your knowing it. (Ph.D., Environmental Science, University of Massachusetts, Boston)

If you love your research area, and we hope you do, then the distinction between work and curiosity/excitement becomes blurred. Perfecting your techniques and acquiring data will not be considered chores, but instead will become a way of life. Your science will fill your thoughts; participating in scientific pursuits will feel natural and will be what you want to be doing. There will be no need to have someone watching over you with clock in hand. Many students are so thrilled by the opportunity to run sophisticated experiments, and are so electrified by the prospect of revealing the workings of nature, that they are scarcely aware of the passing hours.

◆   I worked what I would call long but civilized hours. I normally came in for at least part of the day seven days a week, but went home every night to sleep. Hard work and long hours were expected in my group, but never coerced; basically, I worked hard because I enjoyed it and I wanted to find out the answers and be successful. At that time my research group had rather too many students for the amount of equipment and I found that being a natural morning person was greatly to my advantage. I would come in about 7 a.m. and have almost everything to myself for two hours or so. I did most of my experiments on an apparatus shared mainly with one other student, and we were a perfect match, as he was a dyed-in-the-wool night person while I was the opposite. We often ran the apparatus nearly 24 hours a day, with me taking the early-morning-to-early-evening shift and he the night shift. (Ph.D., Chemistry, University of California, Berkeley)

How Do You Know When You Have Done Enough Work for the Dissertation?

Only in a strange world such as graduate school would one have to ask the question, "How do you know when you're finished?" Surely med students or law students are quite aware of the requirements set down by their programs, and know what they have to do in order to graduate. For Ph.D. students, however, it is not so clear.

◆   There was no clear-cut communication from the advisor as to when I was going to finish and when I was going to write everything up. (Ph.D., Physiology, Kansas State University)

While a scheme of experimentation is laid down at the time of the proposal defense, research results always have a mind of their own. Some experiments may have to be altered or abandoned and replaced when it is clear that they are not working out. Research directions may change—sometimes radically—as the answer to one question naturally brings other questions to mind. Thus the "contract" made with the committee is flexible and allows for some modifications in plans. Even if things do go as anticipated, there is always another study that can be run, another avenue that can be explored, or another analysis that can be performed on the data. There is little "OK, do this and you are going to be out by this time." So, when is enough enough? When does the student think, "That's it; I'm going to write this all up and split"? And when does the committee agree?

Certainly, an extensive and cohesive body of research must be performed before a student can seriously start thinking about finishing up. Yet this is a subjective issue; the minimum amount of research sufficient to constitute a dissertation, particularly when the final research differs from that laid out in the proposal, is a gray area. Without frequent discussions with the committee, it is easy for a student to complete an adequate amount of work without realizing it, and some students actually do much *more* work than necessary. Of course, the opposite scenario also occurs.

♦ I found out that I had done enough work for the dissertation when my advisor told me, around the end of the third year, that I should start thinking about what I wanted to do for postdoctoral work, because as soon as I finished the two projects I was working on I could write everything up and leave. At the time it came as quite a surprise that I might soon be ready to leave, but in fact my advisor was right, because by the time I finished those two projects I was getting a little bored and was ready to move on. (Ph.D., Chemistry, University of California, Berkeley)

♦ I had worked out a general plan of research with my committee at the proposal defense. Some of the experiments worked, and some didn't. I wanted to leave shortly, so my advisor let me replace the experiments that were giving me trouble with some others. He had to fight with one of the committee members to get approval.

I *always* felt like I didn't have quite enough data for a dissertation—right up until the very end. I always felt I needed to do just a few more experiments, replicate something just one more time to make sure it was true. I finally wrote it up because (a) my advisor felt I had enough, and (b) I had been offered a postdoctoral position which I was eager to begin. (Ph.D., Biology, University of Texas, Austin)

It is pretty typical for students to finish up while still feeling that they are "not quite done": there is always more work that they can do or more things that they can learn. But this perception is directly related to the amorphous nature of the requirements for the degree. Accept these feelings, and move on.

♦ Every department has "resident grad students" who have been there forever, and are still there long after you are gone. There are so many easy reasons why they get bogged down (it's surprising that anyone finishes on time in the free-structured research environment!). Their advisor may be to blame for guiding them into experiments where there is nothing to write up after five years. Or perhaps their experiments didn't work, or they got totally negative data—it's not always the student's fault. Sometimes the student is to blame—some are perfectionists, and are never going to finish. They aren't used to the way research works—you get what you get and you don't have to finish up every little loose end in order to write the thesis. "Oh, I only have one more experiment to finish and then it will be perfect." And then it will be, "Oh, I have another experiment to finish and then it will be perfect." You can't be doing that. At some point you have to say "That's that." Get it out! (Ph.D., Psychology, biopsychology research emphasis, Northwestern University)

How Much Independence Is Expected of You?

Upper-level graduate students are expected to be reasonably independent in terms of thought and action. This does not mean that they should fly off completely on their own (a restriction that frustrates some), or consider performing costly experiments without their advisors' prior consent. On their own initiative, however, they may think up and try out new approaches, fiddle with new techniques, and troubleshoot. They usually take care of minor technical problems themselves and do routine equipment maintenance, such as oiling, reloading, cleaning, and simple repairs, without calling in a repairperson. Many students are also very good with their hands, and are able to design and build equipment that is not available commercially. In general, grad students seem to enjoy getting things to work, making things more efficient, fixing things, and just fiddling with things.

As an advanced student, you are expected to be capable of interpreting most of the results of your research, suggesting the logical, next experiments to run, and writing the work up for publication. You should be able to meet deadlines, solve many research problems on your own, make many decisions, and know when to ask for help. Importantly, you must stay busy; you must take responsibility for moving your studies forward with resolve and conviction. Do not wait around for your advisor to tell you what to do at every turn; he may never get around to it. Go after it yourself. Look things up. Speak to other researchers. Learn techniques. Figure it out. You have to develop initiative, exhibit independence, and take an aggressive attitude towards your work.

♦ A lot of independence is expected of you—that's grad school. If you don't push your own project forward, no one else will. Most profs expect their students to take charge, to push and become independent. This is difficult for grad students who haven't learned to be assertive yet. It is something that an advisor should nurture. (Graduate student, Neurobiology and Anatomy, University of Rochester School of Medicine)

♦ A good graduate student is self-starting, motivated, creative, critical, independent, and able to ask interesting questions. In addition, it is important for one to realize, understand, and make use of the fact that most of the time you are your own best critic. (Ph.D., Computer Science, University of Southern California)

Your independence of thought may eventually develop to the point where you find yourself disagreeing with your advisor on some issues. Many students finish graduate school holding beliefs at odds with those of their mentors. This is O.K. It is a natural progression, and is often a sign that the educational system is working.

Is There Time for Life Outside the Lab?

Most people that we interviewed felt that one could indeed find time to have some outside fun while pursuing the Ph.D., although in high-pressure labs this can sometimes prove difficult. Leisure time is more easily attainable during the advanced years, when students are finished with classes and qualifying exams, and when they can schedule their work as they see fit. Even beginning students, though, need exercise and other diversions so that they remain healthy and do not burn out on work.

♦ *Of course* you take the time or make the time to have fun. I still play in the orchestra and still go out. I hate exercising, but it does relieve stress and gives me energy. Keeping your research hours regular also helps—I work mostly from 9 to 6, as if it were a job, and then go home. If you want to finish in *only* four years, then you are not going to have much free time; it is up to you. (M.D./Ph.D. student, Neurobiology and Anatomy, University of Rochester School of Medicine)

♦   From the third year on, I had time to enjoy some of the richness of the Boston area: rehearsal concerts of the Boston Symphony Orchestra, sailing on the Charles, 99-cent lunches at Durgin Park, singing barbershop, and hiking in the White Mountains. (Ph.D., Chemistry, Massachusetts Institute of Technology)

Activities for students will not be limited to individual pursuits—departments and labs organize events, too: department softball/volleyball games, lab picnics and happy hours, Christmas and New Year's parties, birthday lunches at restaurants, and dinners with visiting speakers are all very much a part of graduate school. Many of these events provide opportunities for students and faculty to socialize together, a rare occurrence during undergraduate training, but quite common during the education of graduate students.

Department parties and dinners offer more than social interactions. They provide poverty-stricken grad students with a welcome abundance of free food. This may sound like a joke, but the truth is that many students make every effort to attend such events so that they can eat themselves silly. It is not much fun living on the salary of the typical RA or TA, worrying about every expense, unable to find affordable housing, eating lots of rice and beans. Students learn to take advantage of any money-saving situation that they can.

♦   My lab was in the medical center, and every Wednesday, the M.D.s had a seminar in a room across the hall. There were always extravagant platters of food, and the leftovers when the seminar was over were bountiful—fresh fruit, fancy crackers, whole rounds the size of large pizzas of Camembert and Brie cheeses! It must have cost a fortune. Every week, the building custodian and I divided the remains up and took home as much as we could carry. The food would have been thrown out otherwise. I feasted for days. (Ph.D., university/department withheld upon request)

Students naturally form friendships with other students at the university. This is a benefit of grad school that is more meaningful than you might imagine. Grad students are some of the most interesting and intelligent people that you will ever meet; individu-

alistic, free-thinking, and passionate, they come to the university
from states all over the nation and countries all over the globe. As
such, they are an education in themselves. Most Ph.D.s describe
their socializing with fellow students as a highly broadening and
significant aspect of their lives.

Friends, colleagues, work, fun, social affairs, and scientific
events—all of these become intertwined over the years. Graduate
students are caught in a whirlwind of activities, interactions, and
emotions. There is pain, and there is elation, and overall, life is interesting and good.

♦   My ex-labmate, Margie, and I both have distinct memories of
crying, although we can't recall over what. Margie remembers
having a fight with our advisor, Rob, and throwing some papers on
the floor; I remember fighting with him in the courtyard over the
downtime of my ultrahigh vacuum chamber. Margie recalls running into her lab only to find it filled with fluorine gas. I remember sitting in the "sensory deprivation room" soldering connec-
tors and blasting music at four in the morning. I remember all of
the group outings we had at the beach, too many Margaritas, and
back to work in the evening with more coffee. And all of those
countless group dinners/lunches at that awful Mexican restaurant. I remember staying up until 4 a.m. with Margie and Rob so
we could finish a "Science" manuscript and going for ice cream afterwards at some all-night coffee shop in Beverly Hills. I remember those terrible Beach Boys tapes that another labmate insisted
on playing. Rob left UCLA in my fourth year because Georgia
Tech offered him tenure, but he showed up the night before my
final exit seminar and helped me with my talk. I remember Frank
and Margie staying up all night with me, helping me finish my
dissertation, and going to Starbuck's in the morning for coffee and
more coffee. So, you see, our graduate school experiences were
filled with excitement, anxiety, fun, stress, etc., all of which have
faded with the blessing of time. (Ph.D., Chemistry, University of
California, Los Angeles)

11

Some Additional Aspects of Graduate School Life: Lab Notebooks, Etiquette, Competition, Luck

Lab Notebooks

Scientists who have long been involved in research know from experience that there is nothing as important as taking careful and detailed notes on the experiments they run. New students know this, too, but not from the same perspective. Students often view the practice of keeping a carefully detailed notebook as something one does to appear orderly and productive; a notebook is seen as a dated "list" in essence, an account of what one has worked on. But a notebook should not be looked upon as merely a dusty record of labors past. It is a dynamic and active part of the research process, as important to the interpretation of the data as the data themselves.

♦ It is critical that students do meticulous note keeping. Put down everything, *even if you don't think it is important!* (Ph.D., Physics, University of Colorado)

Never underestimate your ability to forget. It is vast. You may remember every aspect of an experiment right after it is performed—the dose of each drug, the time of day you started, the pH of the buffer, the number of hours the tissue was kept in the refrigerator—but you will not remember for long. Pile a few more experiments on top of this one, and the details all begin to get mixed up. Soon you may forget that certain experiments were ever even performed! This is one of the reasons that you need to write everything down in a notebook, as it is happening, or as soon as possible. Be sure to include instrument calibrations and standard-izations, calculations, and any unusual happenings that may affect the experiment's outcome. You will need a detailed notebook in order to figure out why an experiment did not work, and why it did work. Looking back over your notes, you may find that your experiment contained elements that you did not realize were there, or that you forgot about. Let us say that you did a study that worked the first time, but you could never replicate that original finding. Why did you get interesting results in the first experi-ment, but not in any of the others? A clue may be found by look-ing through your notes. The drug used in the first experiment came from a different lot than the one used in later experiments; there is always the possibility, then, that the two lots were not equivalent. Or perhaps the test tubes from Plastikware, Inc., used in the original study, were more chemically stable than the newly purchased tubes from LabPlastic, Inc., used in the other studies (always record any obvious deviation from an original set-up). Then again, maybe the rats in the later studies were showing stress responses that affected their physiology, a result of having been shipped in only one week before the experiment, whereas the first group of animals had two weeks after shipping to acclimate. Without a notebook in which all these details were written down, you would never know which factors to test to try to pin down the cause of the frustrating discrepancy in the data. (Sometimes the accidental introduction of a factor into an experiment can actually be something positive; new and exciting insights can be gained when the supposedly inconsequential factor is shown to actually have an important effect!)

♦ When I started research, I didn't want to stop during an exper-iment and write neatly in my notebook, so I would scribble notes on paper towels that sat on the table top. I thought I would copy all this into my notebook after the experiment was done, but I was often so tired at the end that I held it off until the next day. By then, I would forget what my scribbles meant (or I couldn't read them), and I would end up with this pile of towels with necessary information that I couldn't use. (Ph.D., Biology, University of Cal-ifornia, San Diego)

A carefully written notebook is necessary if you need raw data for analyses that you did not originally plan to run. Say in your past experiments you collected blood from mice in order to measure the effect of treatment on numbers of lymphocytes. You performed the statistical analysis without consideration of the sex of the animal (males and females were grouped together). You later decide that you would like to know if the treatment affects one sex more than the other. If you were suitably careful, you may have written down the sex of each animal that you drew blood from, and thus may be able to match individual raw data with sex, and perform the analysis that you want.

Your notebooks will prove indispensable for writing journal articles and the dissertation, both of which require very detailed information on experiments. You will look up in your notes to see *exactly* what procedures you performed and how you performed them; what companies your equipment, animals, or reagents were ordered from; the age of your animals, and how they were caged, fed, and handled; the physical parameters of the elements of the experiment, such as mass, time, voltage, temperature, pH, concentration, pressure, velocity, distance, or thickness; the raw or analyzed data obtained, and the conclusions reached. Look at some journal articles mentioning analytical techniques that you will be using to see the parameters one has to record, then record these parameters and any others that may be relevant.

But what is "relevant"? Obviously you can't write down literally *everything* about an experiment, or you would never get done. What is a significant factor in one experiment may be irrelevant in another. The color of the lab wall is probably not going to affect your results if you are doing a geochemistry experiment, and dropping a piece of glassware during a physics experiment on radioactive emissions is probably no big deal. But if you are studying behaviors of hummingbirds caged in the room, the bright red paint on the wall *may* have some influence, and if you are investigating adrenal hormone levels in mice, the crash of glass breaking may indeed startle the animals and affect their physiology. Likewise, a paleontological dig is not going to be affected by the humidity in the air, but histochemical reactions can indeed be modified by unusually high humidity in the laboratory. As a new student, you may be at a loss as to what seemingly extraneous conditions may actually be germane to an experiment at hand. You will get a feel for this over time. You may fail to record the age of your animals, and later be admonished for it. You may read that Dr. Smith could not replicate Dr. Jones' experiment, and Dr. Jones will counter that

Dr. Smith failed to consider the time of day of the study, and thus the influence of biological rhythms. You may overhear postdocs commenting that certain cells stick to glass test tubes but not to plastic ones. You may find out accidentally that a red environment, such as a red wall, increases food intake in hummingbirds.

Scientists can never take into consideration all the possible factors that could influence an experiment, but they do the best that they can. How detailed one's notebook is (or whether it is written in complete sentences or mere phrases), varies from lab to lab and from person to person; if you are not sure how compulsive to be, ask your advisor. (If your research has the potential to result in the granting of a patent, your notebook will be important for the patent process.)

Believe it or not, keeping poor notes can actually result in the loss of a grant. If someone becomes suspicious when they cannot replicate your research, and officially challenges your results, a university or government investigation may take place. You will be required to submit your original notebooks with their listing of procedures, raw data, statistical analyses, and conclusions. Being unable to supply this material (notebooks are required to be kept for years after a project is completed), or supplying inadequate or insufficient records, indicates that you are, at the least, inept as a scientist. Your inability to present a complete set of notes may also bring about accusations that you falsified your data. Either way, you will probably lose your present grant and may be forbidden to apply for other grants in the future. Many lab heads refuse to allow their researchers to take notebooks out of the lab for this reason and may even lock them up, fearing that the notes will be lost or even doctored. In the recent past, lab notebooks were always of the bound type so that pages could not be unethically added or removed for the purpose of altering input. Today, because much data is generated in the form of computer printouts, many scientists ease up on the rules, and use loose-leaf style notebooks that allow their printouts of data and graphs to be easily affixed. Notebooks belong to the lab where the research was conducted; if you want to take your notes with you when you leave the lab, take a *copy* of the original.

Lab Etiquette

If you want to make enemies quickly in the laboratory, just break some of the largely unspoken rules that guide behavior there; the cold-shoulder treatment you will receive will not make you feel very comfortable. And it is possible that no one will even tell you what you did to deserve the festering anger. It is unfortunate, therefore, that many newcomers to a lab are totally unaware of what kinds of rules exist in a laboratory situation, and therefore never think to ask about them. While certainly each lab has its own particular set of do's and don'ts, there are surely common codes of correct behavior that are found across labs. Let us look at some of these.

Personal Work Space

Whether the lab area is large or small, it is best if you confine your research and your test tubes to a select region of the lab bench that does not impinge upon the work space of others. This is not always possible of course, as some equipment areas are meant to be shared and commonly used. However, in general, if it is possible to stick to your own little fenced-in yard, do so. This keeps elbows from knocking, and allows the individual researchers to concentrate on what they are doing.

Personal Equipment/Reagents

Individual lab members may have small pieces of equipment, instruments, reagents, etc., that they can call their own for the duration of the time that they stay in the lab. Since people get used to working with the same particular materials, trusting them and knowing what they can expect from them, it is common courtesy to respect this "ownership" arrangement. Bob is going to be furious if he has to search around the lab to find his heating block, and Karen is not going to be too pleased when she finds that someone has dulled her sharp scissors. You will be amazed how possessive people can be when it comes to their lab equipment. If you ask politely, you can "borrow" stuff for a while, but be sure that you treat someone else's things carefully, and always return them.

A word of caution: never, ever help yourself to someone else's stock or working chemical solutions without asking. If *any* doubt exists, mix up your own.

◆   One night I was working very late. I was hungry and exhausted, and when I saw that I didn't have enough saline solution, I felt it would be OK if I took some from another grad student's bottle that I found in the fridge. I came in early the next morning

to replace what I had taken, but it was not early enough—the student had already run an experiment and had to mix up more
saline. She was burning mad. I felt terrible. It took me a long time
to win back her trust. (Ph.D., Psychology, biopsychology research
emphasis, University of California, Los Angeles)

General Equipment

Most of the equipment in a laboratory is for general use, to be employed throughout the day or week by various lab members. Some,
such as a cryostat or a perfusion setup, may be managed by the use
of sign-up sheets, whereby the researcher signs up in advance to
use the piece for a certain day or a certain hour. Signing this sheet
should be thought out carefully, as others in the lab will adjust
their schedules according to when the equipment is free, and do
not want to find out that a supposed "occupied" time was never
really utilized, and could indeed have been used by them. If you
find that you will not be needing the equipment after you signed
up for it, be sure to cross out your name as soon as you realize so,
not moments before your slot begins. When you are actually
working at the piece, keep a watchful eye on the clock. Pace yourself so that you can finish and still have plenty of time to clean up
and have everything in working order for the next user. You never
want to extend into another's time slot.

What about equipment that does not usually have sign-up
sheets, such as scales, pH meters, short-term centrifuges, etc.?
These, too, have rules associated with them. First of all, do not hog
equipment; do your work within a reasonable amount of time, and
get off the piece the minute you no longer need it. If you are working on a large-scale project and have been using the equipment for
a long time, and you see others waiting, it would be polite to let
them squeeze in if possible. Never leave your work half done. Do
not leave some weighed chemicals sitting on the scale while you
go do something else, and do not leave test tubes sitting in the cen-
trifuge when the spin is over. Finally, even though a piece of equipment is about to be used by another person, you should leave it
clean, reset to the starting position, and (often) shut down.

Different Work Styles

Individual researchers all have their own, unique ways of doing
things. They may have been trained by different teachers, or they

may have just developed their own styles or techniques for performing procedures. They all have different paces of working, too. Try to respect this.

The Last Spoonful

Each lab has its own ways of dealing with supplies that are running low. If a student uses up the last bit of a stock solution that is used commonly by all lab members, he may be responsible for mixing up a new batch, or for informing the lab technician of the situation. If someone notices that the supply of magnesium chloride is running low, he may be responsible for ordering some more or for informing the technician to order more. This is usually done *before* the bottle is totally empty, as it takes some time to restock.

Labeling

Every substance in a laboratory is labeled; there should be no need for guesswork. If you prepare some reagent, you must immediately label the bottle with your name, that day's date, and the specifics concerning the contents within. This is so basic that people will be shocked if you fail to comply.

Computer Fiddling

People who sit down at the computer do not like to be greeted with surprises. So, do not change the way the lab computer is organized unless you OK it with the rest of the lab first. Be sure not to delete files that you never use; someone else may be relying on the information they contain.

The Technician Is Not Yours

You will probably discover that the technician is the one person in the lab who knows everything about how the lab works. While the tech may or may not know much about the lab's research efforts, he will undoubtedly impress you with technical know-how, managerial skills, and record keeping. While technicians can certainly make a student's life easier, by performing a variety of technical duties (such as preparing reagents commonly used by lab members), organizing the lab, or ordering supplies, keep in mind that they do not work for you. They perform *general* duties and, unless the lab head has assigned them to specifically help you on a project, they are not for you to boss around. It is likely that you will be expected to do much—if not most—of your own technical work.

Noise Warnings

In many labs, use of some pieces of equipment (opening a carbon dioxide tank to create dry ice, for instance) produces a sudden and loud noise. This sound is startling and upsetting to those who are in the lab and unprepared for it. It is customary to therefore yell "Noise!" or some other prearranged warning, before the equipment is used.

Favors

It is perfectly O.K. to ask your fellow lab members to help you out now and then—to take your slides out of the oven if you will not be around, to assist you in a surgical procedure, to help you with a mathematical problem, to fix the oscilloscope, to teach you how to run a gel. Getting assistance from other grad students and post-docs is one of the main ways you learn, and "I'll help you today, you help me tomorrow" is the way many a heavy-duty experiment gets done. Just do not ask for an excessive number of favors. *You are expected to do your own work*, or at least give it your very best try. Do not exploit your labmates, or you will not get help when you really need it.

"Your Mother Doesn't Work Here."

This saying, written out on a card or a piece of paper, can be seen hanging above the sinks and lab benches of many a lab. "Clean up after yourself" sometimes precedes it, and clarifies its meaning. Cleanliness and neatness are high priorities in most science laboratories. So heed the sign, wash your glassware, put away supplies and instruments, wipe off the bench top, and leave the lab the way you found it.

Safe Handling of Dangerous Substances

Most biology and physical science laboratories contain chemicals/biological materials that are biohazardous, equipment that uses dangerously high voltages, or instruments that can cause physical injury. There are proper ways to handle all of these, and if you are not taught what they are, ask; improper handling puts all lab members at risk. If you are found continuously careless in your habits, you are going to be asked to leave the lab.

Simple Respect

There are innumerable other situations in the laboratory that demand propriety and, although the specific behavior required varies from case to case, a good rule of thumb when in doubt is simply to treat the other lab members with respect. The sometimes casual atmosphere of a laboratory can be deceiving; serious work goes on there. If someone is obviously concentrating hard on their research, do not stroll over to chat or beg a favor. If someone is waiting for you to finish using a piece of shared equipment, do not dawdle. If you use the vibratome, turn off the light when you are finished. If others do not like the radio playing loud music, do not turn it up. Respect the time, efforts, and property of others. And maintain a working relationship with your lab members, even if you do not particularly care for them.

♦   I had no serious problems interacting with others in the lab. There was often a lot of competition for equipment and occasional anger at someone breaking something and not fixing it, but for the most part we respected each others' rights. The group mostly policed itself but our advisor kept his eye on things and intervened whenever a "higher authority" was needed. (Ph.D., department withheld upon request, University of California, Berkeley)

♦   The lab I was in sometimes resembled a three-ring circus! I didn't have problems interacting with anyone in my lab because if I did I confronted the situation and got it straightened out. Thus I became the group social worker and everyone wanted to tell me their problems. All of the people were nice one-on-one but the mixture was caustic! My advice to avoid this situation is IF you have a problem with someone, then talk to them about it. It does not have to be an argument and nine out of ten times you will feel better and things will be fine. The alternative is to dwell on the situation, create negative scenarios in your head and bad-mouth the person. All of those things tend to make the whole situation feel overwhelming and then it interferes with your research and life outside the lab. (Ph.D., university/department withheld upon request)

Competition

Competition among graduate students can range from low to high levels—as graduate school is no different from any other place in this regard. Whenever people are being graded, evaluated, and judged, a certain amount of jealousy and resentment is likely to circulate among them.

♦ I don't know if you can really label it competition, but everyone is pressed to pass qualifying exams, publish papers, etc. I don't know if one is fully aware of it or not, but one is always comparing oneself to other students in terms of passing these landmarks. How come this person has already finished all her course requirements and I haven't? Where do I fit in? It's hard to handle. But this is true in all aspects of life. (Graduate student, Neuroscience, University of Rochester)

♦ You want to be recognized. You want to know, am I doing something worthwhile? And all of this is *relative to others.* (Graduate student, Psychology, biopsychology research emphasis, Rutgers University)

Competition may be at its greatest level during the class-taking years, when exams and grades make it easy to compare people. Later on, in the lab, competition may creep in if one particular student is making most of the significant discoveries; this can be hard on the remaining students, who may feel that they are useless or not contributing enough.

♦ The students in my entering class were pretty competitive. They would brag that they had no trouble doing the problem sets, and then you would find out that they had been to the library and had found the answers in a book. (Ph.D., Physics, University of Illinois)

People are people, and there are always going to be some individuals with a strong sense of rivalry, perhaps even a few who might go so far as to cheat. This is partly fueled, no doubt, by the competitive nature of the general scientific arena; the quest for recognition, the race to "publish first," and the tight job market found in this arena. It must be noted, however, that due to the nature of graduate programs, competition in graduate school generally tends to be *minimized*. This cushioning of the competitive spirit stems from two sources. First, graduate education is not by any means a "group" experience; that is, it is highly geared to the individual and to his or her likes, abilities, and personality. The number of entering students in individual departments is usually small and, except for some common classes, students usually disperse very quickly to different labs that appeal to them and that agree with their natures in terms of subject matter, technical proclivities, working hours, etc. It is not very common for two students from the same entering year (the students who might likely compete with each other) to be in the same lab. Thus, unlike the situation in medical school, where the cohesiveness of each class is maintained, in graduate school the identity of each class is blurred. Also, students work on very different projects, and they go through graduate school at different paces. They reach the different stages of the program at different times, and they can vary in the time it takes to finish by a year or more (this usually says nothing about the student—some students do more experiments than others, or their experiments require more time to do). Because students in a department specialize in different areas, they are also not usually competing for the same postdoctoral position or other job. All of this tends to minimize competition in its usual sense; it is more common to have students in a given lab compete for equipment, space, and their advisor's time.

The second reason for the relatively small amount of rivalry in graduate school (as compared to many other institutions) stems from the generally liberal attitude of academia. Here, individuality is accepted, and even flourishes. Personal idiosyncrasies are overlooked as long as thinking is logical and research is competent; graduate students are free to be themselves, to follow their own interests, to pursue their own talents, to let their minds wander freely and unfettered. Each student is on his own trip, his own search. Competitiveness is not entirely consistent with such gratifying individualism.

♦ No one else was working on the project that I worked on, so I didn't feel much competition. Other graduate students in the lab were very helpful. About the only competition was for scarce supplies or resources—for example, who got to use the choicest recording equipment. (Ph.D., Psychology, biopsychology research emphasis, Cornell University)

The Role of Luck

Luck plays more of a role in scientific discoveries than most scientists would like to admit.

There are two types of "lucky" findings that warrant discussion here. One type occurs when an experiment merely "works." That is, data are obtained that can be published: an effect is definitely noted under certain experimental conditions, or a clear characterization of certain properties is successfully obtained. These findings are considered lucky only because many experiments do *not* work out—they have technical problems, or they yield data that are either uninterpretable or show no effect. This can happen in even the most carefully and intelligently planned study. Thus, it is an event for celebration when a student finds an experiment that produces results that can eventually be published; this is even more spectacular when the experiment "works" after only one or a few trials, needing little or no change in experimental design or procedure. If the student is very lucky, he or she may find a string of experiments that work, seeding a number of published papers. Of course, luck is not the only element operating here; the more one learns, thinks, and experiments, the more proficient one becomes and the more ideas one generates, and thus the greater one's chances of designing and performing experiments that turn out successfully. The latter is only one scenario among many, however. Some unfortunate students devise experiments that, for all their cleverness and creativity, and despite being approached, designed, and executed appropriately, never "work." Due to no fault of their own, except their unusual misfortune, these students must accept their degrees based on their efforts and not their productivity, a terrible state of affairs in this "publish or perish" business.

A second type of lucky finding that happens in research is the kind that is made quite serendipitously while in pursuit of an unrelated quest. A student may just happen to notice, for instance, during a study on weight loss, that rats fed a diet rich in vitamin Z actually live longer. While this finding (a relationship between vitamin Z in the diet and longevity) has little to do with the experiment at hand (effect of vitamin Z intake on weight loss), it is recognized as being significant, and deemed worthy of pursuit. Besides luck, it takes an educated, prepared, and clever eye to recognize the observation for what it is, and not disregard it. A finding like this can lead a student off on a path of her own, separate in orientation from that of her advisor. If the finding is of great significance, and if the student makes the most of it, it could make the student's career and she could easily spend the rest of her working years exploring the discovery. But such findings are not very common, and students really do not have to go off in totally different directions from the rest of the lab to be successful. The results of their quite "ordinary" experiments will lead them down unpredictable paths and, with any "luck," they will find something interesting—even within their advisors' projects—to call their very own.

◆　I had what I would consider nearly an ideal journey through graduate school. I was given a project that provided the opportunity for me to do some really fascinating, beautiful, and original science that combined theoretical and experimental work in a very satisfying way. In the end just about everything I tried to do produced useful results, although some things worked better than others, and there were some periods of weeks or months of considerable frustration. Unfortunately I can't say how I was able to arrange this; being fundamentally smart, sensible, and willing to work hard certainly helped, but part of it was pure luck. I'm convinced that a talented person with a good attitude will nearly always do "okay" in graduate school, but I also know that many really good people (who proved it by their successes later on) had a rather stressful and generally unsatisfying time in graduate school. (Ph.D., Chemistry, University of California, Berkeley)

12

Do I Belong Here?:
Insecurity and Stress

Grad school engenders insecurity simply by the way it is run.

The research enterprise at a university is a massive operation with an administrative core, a conglomerate of dedicated personnel, vast capital, and ambitious plans. Although it springs from the combined efforts of individual researchers, it transcends any one individual or individual lab; it certainly does not start or stop with each new group of graduate students, and graduate education is not its only, or even main, concern. Grad students are not kept separate from the heart of the enterprise; they are not trained in isolated chambers until they are up to snuff. They are dropped into the research environment and they learn on the job: they participate in or attend key activities such as journal clubs and seminars right from the start, and many are expected to perform experiments in the lab or field shortly after their arrival. It is only in classes that students are solely with others at their own level. Otherwise, they are thrown in with more advanced students, postdocs, and professors, and are expected to blend in and more or less keep up. There is a recognized disparity between new students and their more advanced colleagues, yet people *act* as if there is not any. Sure, beginners are taught techniques in the lab or field, but they are treated as professionals, and are expected to pick things up rather quickly so as not to slow down the research process. At journal club or seminars, topics are not geared to their level and no attempt is made to explain those things that a newcomer assuredly does not understand. Graduate school is like an automatic revolving door: the student steps inside for awhile and is forced by the motion to keep up with the pace. The door does not accommodate to the stride of the occupant and, likewise, the churning of the research engine does not slacken for the uninitiated pupil.

As a result, it is easy for a beginning grad student (especially one whose undergraduate degree is in another field) to imagine that he or she is woefully unprepared for, or incapable of succeeding in, graduate school. When things are often over one's head, this is a natural way to feel!

♦   I initially found journal clubs frightening. I had no idea what a journal club was before I went to graduate school. I was pretty much in a daze. People were using language and abbreviations that I didn't understand. I just sorta sat back and thought, what are they talking about? It was pretty overwhelming. (Ph.D., Biological Sciences, University of Texas, Austin)

♦   In undergraduate courses, the professors assume that you don't know the subject, and they start with easy material and work up to the complicated. When I began some graduate classes, the professors just started talking away, assuming that you already had a certain level of knowledge. I thought: What is this guy talking about? Professors also assign journal articles, and if you haven't read scientific papers before, it can be really tough. In addition, grad students tend to be shy in class—they're afraid to ask stupid questions. (Graduate student, Psychology, biopsychology research emphasis, Rutgers University)

Beginning grad students may exaggerate the background knowledge and capabilities that are expected of them. Many worry that they are going to be "found out." Professors do indeed have high expectations of new students, but their expectations are also reasonable. They realize that these novices do not know as much as those who have been around longer (although every grad student comes in with a different amount of knowledge, and some do know much more than others). The assumption is not that new students can comprehend all they hear; the assumption is that they can learn quite a bit over time by exposure. "Green" students are expected to watch and listen and to pick up whatever they can. Not understanding what more advanced researchers are saying to each other should not be a cause for alarm.

♦   When I first entered graduate school, I was concerned that the department would figure out that I was not as smart as the other graduate students. I wondered what kind of loophole I fell through to get into what I thought was a prestigious program. If they kicked me out, what was I going to do? Everyone seemed to know exactly what they wanted to study and who they were going to work with. The seminars I went to that first year fueled my feelings of inadequacy; they were either too sophisticated for my level of understanding or on topics that I did not understand. It took me at least a year to understand that I belonged, and to accept that I could not be (nor was I expected to be) an expert on every topic.

Whenever I felt insecure in grad school, what made me feel better was remembering what I overheard a visiting professor say: "Postdocs don't know very much—they're still students." From that, I realized that science is a continuous learning process, and that everyone, even if they already have their degree, still has much to learn. (Ph.D., Cell Biology, Baylor University)

♦   It's a gradual process. At first I didn't understand much of what was said at seminars or journal club, although I didn't mention it. After some time, you think, oh, maybe I understood *this* one, so it isn't so bad. Much later, you think, well, if I didn't understand, so what, it's not important for my career; I realized that I didn't have to understand everything that was not in my area. I gathered that no one was going to ask me, as a first-semester grad student, my profound theories or opinions on an issue. It was assumed that we were there to assimilate. I almost never talked in journal club. The only time I spoke was when I was required to do so. That was scary; I hated it. (Ph.D., Biological Sciences, University of Texas, Austin)

♦   I attended one of the top graduate schools in the country, and nearly everyone in my entering class had been at or near the top of his class as an undergraduate. Clearly we couldn't all be at the top in graduate school too. (Ph.D., Chemistry, University of California, Berkeley)

Since it is not clear what you should know, and what you do not yet have to know, it may be best if you just keep quiet and listen until you get up to speed (this does not apply to classes or your research, where you definitely should ask questions if you do not understand). Silence can easily be mistaken for knowledge, so others in the room are likely to assume that you understand what is going on. Just let them assume this—it cannot hurt. With time, you will understand the jargon and things will make more sense. You certainly won't understand *everything* you listen to but, then, no one else will either.

Classes and the qualifying exam are another source of anxiety for many in a Ph.D. program. The "quals," in particular, are met with tremendous trepidation in many departments. Unfortunately, the stresses involved in becoming a Ph.D. candidate cannot be avoided, although joining/forming a student support group or a student study group can be helpful in relieving tension. A performance in classes or on quals that is mediocre may fill a student with self-doubt; the student may even seriously consider dropping out of the program. If science is what he really wants, however, and if the department is willing to keep him on, the student's past should not hold him back. Many students really "take off" when they finally start doing experiments. Things finally "fall into place" for them in the context of the lab or the field. There are many university professors who were late bloomers.

♦   Getting a Ph.D. in chemistry was a difficult, rewarding, and enjoyable experience that I treasured then and now. Coming from a small liberal arts college put me at a disadvantage in science courses for the first two years, but once I got past the hurdles of qualifying exams and into the lab, I started a hot project, which formed the basis for several papers, and a patent. (Ph.D., Chemistry, MIT)

♦   There were so many sources of stress that it is hard to know where to start. The primary ones were lack of time and perpetual shortage of money. Anxiety over exams (one written and two oral) was also high on the list. In addition, my undergraduate years were at a small liberal arts college, so grad school at a large university was stressful in that I had not had access to the more specialized courses that the other students had already taken. Since these subjects were covered in the written qualifying exam, I had a lot of catching up to do. Marrying during graduate school was another source of stress—partly because of my choice of husband (!) but also because it was difficult to give sufficient attention to both husband and research. (Ph.D., Zoology, University of California, Los Angeles)

♦   Feelings of insecurity and stress were my worst obstacle. My solution was to stop thinking of my advisor as an evaluator, and to begin thinking about him as a facilitator and resource. My saving grace was strong peer-group support. Get to know the other students in the department. Get involved in student organizations. See that your lot is not unique. (Ph.D., Physics, Princeton University)

Another storm of anxiety can arise over the prospect of picking an advisor (i.e., pinpointing your research interests and narrowing down your choices of researchers). It can also be difficult getting a potential advisor to take you on.

♦   My biggest stress in grad school was getting past the qualifiers and then finding an advisor that I wanted to work for. I wanted to be a theorist, but my chairman kept telling me that I should be an experimentalist because I didn't do theory fast enough. (I always feel like I'm stupid—some people can solve in an hour a problem that takes me four hours to solve.) Once I hooked up with my advisor—he was wonderful, and a first-rate scientist and human being—I felt extremely fortunate. (Ph.D., Theoretical Physics, Brown University)

♦   I was very interested in cross-disciplinary research, and switched graduate departments from physics to biology. I was never discouraged, but I *was* looked at with quite a bit of skepticism ("But, that's not PHYSICS!/BIOLOGY!"). From a scientific standpoint, I was quite isolated from fellow students. Because neither department knew what to make of me, I was not allowed to carry credits from my old department. This, combined with having to take background courses, lost me at least two years toward my degree. The first year was *extremely* tough, both academically and emotionally, from culture shock and from the lack of a scientific support group. I seriously considered career options outside of grad school, but after discussing it with a mentor, I made a list of *ideal* goals I had when I chose the road I did. I gave grad school another year to see if some of these goals were reachable. Most importantly, could I simply get back to the science, *do* it, and enjoy it? This removed the onus of *performance* from me. I stopped worrying about impressing everybody (to some extent, anyway) and the debilitating angst about *everything* was lessened, and I concentrated on trying to enjoy the actual research.

Once I started getting out to present my work, I was overjoyed to discover the larger scientific community. Happily there are always the "enlightened" few (i.e., those that think like I do) who are excited by cross-disciplinary research. The colleagues I encountered, from grad students to division chairmen at NIH, were supportive and enthusiastic. I was excited by their work, and got to have my work critiqued by an informed audience. (Ph.D., Biophysics, Princeton University)

Finding a narrow area of research for the dissertation, and a specific project(s) to work on, is nerve-wracking and frustrating. Tension and uneasiness always accompany the precarious and rootless period before a dissertation project is securely in hand.

This time can also be one of self-doubt. Why am I finding it so hard to get pilot experiments to work? Am I capable?

♦ My biggest challenge was figuring out what I was interested in, where my questions lay, and thus what I was going to work on. One way to explore is to *just start working* on some project. (Graduate student, Neuroscience, University of Michigan)

♦ I have days where I find myself thinking that I am not smart enough to think up experiments, or that I could never be a "real scientist." I have had a number of experiments that didn't turn out the way I had expected or hoped and I find that it is very easy to start thinking that somehow it is my fault—that if I were a "better person," or worked harder or was smarter, I would have been able to get more exciting data. I never get praise for my efforts or positive feedback, so it is difficult to feel good about myself. I don't receive negative feedback, but I don't get much encouragement and support. You need to learn how to find your own sources of encouragement in grad school. And you need to learn how to counsel *yourself*, that, no matter what your data show, you are still obtaining valuable experience in how to think about experiments, how to perform various techniques, how to ask questions, how to work with people, all of which you need in the long process of becoming a scientist. (M.D. and graduate student, Neuroscience, New York University)

Ph.D. *candidates*, past coursework and quals and heavily involved in research, continue to face stresses of many kinds. They endure monotonous procedures, suffer frustration over research problems, and may feel lonely and isolated or overwhelmed by the amount of work. They suffer insecurities about abilities and productivity (especially if they do not receive feedback from their advisors), and crave approval that may never come. They are stressed in high-pressure labs, and made uneasy by the huge egos they confront there. They fret about not having enough money, and hate the restrictions that being poor places on their lives. They worry about getting a good postdoctoral or faculty position. Not every student experiences these stresses, but most will experience at least some of them at some point. If these normal anxieties and pressures become unbearable, the student should seriously consider taking advantage of the (usually free) services of the campus psychological counseling office. Contrary to what you may think, graduate students are one of the most common groups of students using these services.

♦   Stress, insecurity, depression: there has been a lot of all that. For years I was insecure that I wasn't good enough to actually get this degree. When I started my project, there was no guarantee that it would even work, so failed experiments were very stressful. Now I have to find a job, and that brings its own stress and insecurities. And yes, I frequently get depressed that I am living in poverty, doing difficult work with what seem like bleak job prospects for the future. (Graduate student, Biophysics, University of Rochester)

♦   You read and read and read, and you think, there is *so* much out there—I can't possibly synthesize it all. I even find it difficult to examine scientific papers effectively (grad schools should have a course in this)—I have sat next to experts who say, "This method is crud, the statistics are wrong, the results don't make sense," and I'm sitting there thinking, "Why didn't I catch any of that? I thought the paper was just fine."

I also am insecure about the quality of my own work. I seem to need appraisal by an authority in order to judge what I do. Fortunately, one gains more and more confidence in their work as they go through the graduate program. (Graduate student, Biology, New York University)

♦   There are always more papers to read than I can possibly read, and more work to be done than I can do. I haven't yet learned the art of being satisfied with what I do, in fact, accomplish—I seem to always feel guilty that I "should" do more. So, I always seem to have an underlying feeling that I'm "not quite good enough." I think this feeling stems from the nature of life itself and is probably *not* because of a unique quality of graduate school—thus, I view graduate school as an opportunity to learn ways to either accept this constant feeling or change it.

Another, perhaps trivial, source of anxiety stems from sharing a small office with another student. I work best in a quiet atmosphere where I can be undisturbed. I therefore spend little time in my office. I either go to the library to read, write, or study, or I go home to my computer. I worry that since people do not see me in my office, they will think I am not working, especially since I have observed that the people who are "always around" are praised as "hard workers." I feel I put in *at least* as much time as these "hard workers," but since I am not as visible, I do not receive this praise. (M.D., graduate student, department withheld upon request, University of Michigan)

♦   I'd like to talk about . . . the loneliness and isolation graduate students go through. This is a function of several factors—the number of people in one's program, the nature of the work one

does, how well one is doing, one's relations with one's advisor/
professors, etc. . . . Grad school in general, and research in particular, are very isolating experiences. Research is a very individualistic activity. (Graduate student, department withheld upon request, Emory University)

♦ They tell us that research is 95 percent failure, and only 5 percent success. We work and work, and it is very hard to find hope
in the 5 percent. When things don't go well, it's discouraging. I immediately feel guilty for wasting money, and in my type of work,
an experiment can cost $1,500. (Graduate student, Neuroscience,
University of Rochester)

♦ A big challenge for me, even more so than tests and orals, was
figuring out what the data (all those numbers and points) meant,
and how they should be interpreted. I found this very difficult.

   The most stressful aspect of graduate school was the putting
of lots of time and thought into projects which did not work out,
and not being able to determine WHY. I spent over a year on a project which ultimately could not answer the question I was asking;
to this day I cannot say whether the technique was inadequate or
whether I got a negative result. But I think you have to remember
that whatever results you get, they can be part of your "story."
Whatever results you report can help someone else pick up where
you left off. You don't have to cover every angle, explain every
contradiction, have everything fit neatly into your hypothesis. Remember, you are only working on one tiny bit of the puzzle.
(Ph.D., Biological Sciences, University of Texas, Austin)

♦ You really have to know when to stand up to your advisor.
One student I knew worked on a theoretical model for a whole
year with no results. Finally, he got the nerve to tell his advisor he
was going to drop it because it wasn't going anywhere. (Ph.D.,
Physics, Brown University)

♦ Feelings of insecurity, anxiety, or depression in grad school
are totally normal (I hope!). Don't sweat it. (Ph.D., Biology, Yale
University)

Dropping Out

Some grad students were just not meant to be scientists, and they
usually sense it. Little by little, it dawns on them. It may be that
they found courses a huge struggle, they did not do well on the
qualifying exam, and they find that they do not have a knack for

the practice of research, either. Or, as the years pass by, they realize that nothing is really catching their interest, and they find it hard to concentrate on their work. Perhaps they grow interested in other professions, finding them more exciting than science, and they begin to recognize that somehow they have made a big mistake. Many grad students never make it through graduate school, and most of these, not wanting to continue or sensing inadequacy, leave of their own accord. Forty percent or more of grad students in the sciences drop out before getting their Ph.D.s. (Bowen and Rudenstine, 1992). For those whose abilities or interests lie elsewhere, dropping out is often the right decision.

Students who complete certain requirements (such as coursework), and then drop out or are asked to leave, are frequently awarded a master's degree as a "consolation prize." This allows them to compete for jobs that require a degree beyond the B.S.

◆ Some students never made it through grad school. Few were actually asked to leave; some, even though they were doing okay, simply decided that they didn't want a Ph.D. in chemistry after all, while others left of their own accord because they knew they were doing poorly. Some of these people ended up going to grad school elsewhere, earned Ph.D.s and were in some cases quite successful later on, indicating that maybe they just needed a little extra help or time. (Ph.D., Chemistry, University of California, Berkeley)

◆ I always maintained my enthusiasm in grad school. If I didn't enjoy what I was doing then why would I have stayed? There are so many people that hate grad school and yet they stay. Why?? Life is too short. So what if you decide it is not for you and you leave—it is better to be happy than miserable! (Ph.D., Chemistry, University of Rochester)

13

Foreign Students: Unique Problems and Stresses

Foreign graduate students suffer all the difficulties and stresses experienced by other grad students—plus many others. In addition to the rigors of graduate school, many international students must also deal with language problems, cultural differences, unfamiliarity with the structure of American institutions, loneliness, and homesickness.

♦ The biggest hurdle I faced was the culture difference and the fact that I got very lonely at times. I kept comparing the U.S. to the U.K., and in the end decided that the U.S. was just "different." The absence of familiar crutches made "downswings" in mood much deeper. (Ph.D., citizen of the United Kingdom)

If the above former grad student, citizen of an English-speaking country, felt lonely and estranged in the U.S., imagine what it must be like for those who are not proficient in the English language!

♦ Like other foreign students, my biggest problems are language and culture. I did not have much time to study English before I came here and I never spoke English in my daily activities. The differences in culture make me nervous—I worry that I might do something that in the American tradition is considered wrong. It is very hard work for me to adjust.

Another problem is that I have no idea about what graduate study is like. I have an M.D. from Thailand. Being here as a graduate student is completely different from studying in medicine.

Everything is very new for me and I am learning all the time.
(Graduate student, citizen of Thailand)

Problems with the English language are the most widespread
difficulties faced by foreign grad students. If English is not your
native language, you may have trouble understanding spoken and
written words, and it will probably take you longer to comprehend
topics discussed in class, to write papers and other class assignments, and to read and understand journal articles and other material. Since the pace of the American classroom is very fast, and
you are expected to keep up with what is taught and assigned, you
can easily fall behind very quickly. Importantly, you may also find
that you have less time for research.

If you find yourself having problems with language, we recommend that you take a course in English; English as a Second Language (ESL) teachers on campus will be able to suggest a course
suited to your level of knowledge. Even students who consider
themselves quite proficient in English may experience difficulties
understanding "jargon." This can be very frustrating, especially if
you came to the U.S. confident in your language abilities. If you do
not understand the meaning of something you hear, do not be em-
barrassed to ask the speaker to explain it to you.

> ♦ I have been in the U.S. for six years, and the difficulties
> change year by year. For my first year, it was a total culture shock
> and a language problem. Before I came to the U.S., I *did* learn En-
> glish and things about the culture. However, what I experienced
> here was totally different from what I learned. Too much slang
> and segmented sentences are used in American English. It took
> me three years to get used to it. In the American classroom, teach-
> ers speak the formal and correct English, so it is not a big problem
> for me to understand lectures at all. I do have difficulties giving
> seminars and oral presentations. This probably results from the
> language problem. In addition, the thinking logic for people in the
> U.S. is different from that in my country. This made it a little dif-
> ficult in the beginning to communicate with people. (M.D., gradu-
> ate student, citizen of Korea)

Another frustration you may experience involves the difficulties *others* have understanding *you.* If you have a strong accent,
people may be unable to comprehend what you are saying, and
they may frequently ask you to repeat yourself. You may find this
embarrassing or annoying. Foreign students who have difficulty

understanding others, or making themselves understood by others, may worry that professors and other persons in the department see them as "stupid." They may grow more and more insecure, especially if they are making errors which disturb or humiliate them. This is especially painful considering that many foreign graduate students were the top students in their country. It is probably a figment of the imagination of international students that American professors are thinking poorly of them. However, because language and cultural differences are obstacles to communication, it is difficult for professors to gain information about these pupils; they represent "unknown entities." For this reason, it is especially important that foreign students resist any urge to "fade into the shadows." Instead, they should strive to make themselves known, by conversing regularly with faculty members and keeping their advisors informed about the progress of their work.

♦   In my opinion, the biggest challenges of a graduate student besides doing science are public relations, public speaking, and scientific writing. You have to establish trust, demonstrate that you are capable of doing experiments and are a critical analyzer. I put myself in the shoes of my advisor, and thought about what I would expect from a student. This helped me to take his criticisms in a positive way (not always though). I learned that it is very critical to inform your advisor about the progress that has been made in the experiments. If you don't discuss this with your advisor, he/she may think that you are not doing anything. It is also essential to talk about your research with others in the department. This helps in the formulation of new ideas and makes us think globally instead of in an isolated way.

After coming here, I found that life here is totally different. It took me some years to understand how the system operates. This was a stressful period and sometimes interfered with my working capabilities. Then, I decided that I should get out of this mode and do something about finishing the program. I started working on four different experiments at the same time, and kept a schedule for everything I had to do. I had to do this as this was the best solution I could think of to get the Ph.D. (D.V.M./Ph.D., citizen of India)

International students are likely to find the American educational system to be very different from the educational system they are used to. In Japan, for instance, it is very difficult to get into a university, but once a student is accepted, staying in is easy;

Japanese students may thus find the workload, and the amount of
testing and evaluating in American graduate schools surprisingly
excessive. Furthermore, American students and professors tend to
interact in an informal and casual manner. Rather than "lecturing"
and telling students what to learn, professors often "exchange
ideas" with students. Grad students ask questions and make comments in class and usually call their professors by their first
names; students and professors may even socialize together. Students from different cultures may find this informality uncomfortable, being used to more polite and respectful relationships
with authority figures. Some therefore may avoid participating in
class discussions, asking questions, or interacting with faculty.
This is not good. Speaking up in class if you have something
significant to say or ask, and mingling with faculty members outside of class promotes friendship, shows enthusiasm, and aids
learning.

♦   American society expects people to be assertive, and grad
school is part of that society. I agree that taking charge makes a
good impression in terms of one having initiative, but it is very
stressful. It puts a lot of pressure on an individual whose personality was shaped in a culture where assertiveness is not of such importance. (Graduate student, citizen of Bulgaria)

♦   Professors here are much more approachable. You socialize
with them, and address them by their first names. There are different communication styles between cultures, and I don't think
each side necessarily knows where the other is coming from. Definitely, the working atmosphere is totally different from that in
my country. Independent thinking, encouragement and appreciation of good science are some of the characteristic features of the
scientific community in this country. (Ph.D., citizen of India)

♦   As a foreign student, I find that in a lot of instances I am un-
derestimated, until whoever is doing the underestimating finds
out that I did my undergraduate education here. It makes me wonder why people react this way. There is no reason to patronize
anyone who comes from a different country.
    It is hard for me, with the strict up-bringing that I had, to call
a professor by his/her first name. I had to explain each time that I
just could not do it or that I would require time to adjust. As a result of not being on a first-name basis with professors, I am perceived as shy and lacking in confidence. I will quote an unnamed
person who said that I "am not a go-getter." I took a close look at
myself after I heard that and I realized that it is a cultural differ-

ence. In the U.S., I am expected to show my interest in a particular subject by talking about it at all times even though I am not sure if what I am saying is indeed correct. I approach things differently. I don't think I have to talk about my interests at all times to prove that I like them, or ask questions when I know the answer. This has been one of the hardest things for me in graduate school. I may look shy and not confident (although I don't think that is an accurate description of me), but I know what I want. It's just that I don't express it in the conventional way. (Graduate student, citizen of Cameroon)

In some countries, such as Japan, students work on topics picked by their advisors, memorization is stressed more than problem solving, and challenging ideas or debating with a professor is antithetical to the hierarchical social structure (Normile and Kinoshita, 1996). Students coming from such countries will find the American graduate program's emphasis on independent thinking, creative problem-solving, and liberty to question or disagree with authority, unfamiliar and challenging. They may have difficulties dealing with the freedom to make decisions and to make up their own minds. They may find it a struggle to design their own experiments or to take responsibility for the progression of their research. They may have a hard time doing things independently, and not as part of a group.

♦ The difficulties I have are mostly related with language, such like difficulty in the presentation of seminars, or sometimes difficulties with my own experiments. The other difficulty is that I have to constantly speak up. Many times it is hard. And also it is hard to decide what to do when told "You can do whatever you want." (Graduate student, citizen of Korea)

♦ When I first came here, everything was new to me. I did not have problems understanding what professors said in class, probably because I passed the GRE and TOEFL in China. We have lots of Chinese students in our department, and most are very kind to me, and some are very close to me. They helped me a lot when I first came here. One thing that I had trouble adjusting to was that there are so many choices to make here. Usually it is not clear whether choice A is better than choice B. So, sometimes it is really hard for me to make a decision. The other thing that I had trouble adjusting to is that everyone here is very independent. No one will interfere with your personal life, and no one will tell you that you should do this and not do that. This is very different from the situation in China. In the U.S., everyone is very busy, and you must take responsibility for yourself. We have a very close Chinese student circle here. It functions like a buffer between a foreign student and the American society. This is why I didn't feel great culture shock here. (Graduate student, citizen of the People's Republic of China)

There are other aspects to American graduate programs that may surprise foreign students. Professors may ask questions of students in class, so students must pay attention and keep up with the work covered. Also, each course is not going to have a single, major exam at the end of the year as is common in some countries. If the course has exams, there are usually going to be two or more of them, usually written. A common schedule is to have an exam about halfway through the course (the midterm exam), and another at the end of the course (the final), although some minor exams may be administered during the term in addition to these. One of the authors remembers one graduate student from England who got up in the middle of a midterm exam in statistics and simply walked out of the room; in England, all his exams were oral, and he wrongly surmised that this written exam was a practice test!

Some degree of culture shock hits every foreign student. You are no longer home, and it is very obvious. The public transportation systems are different, as are the phones, the traffic regulations, the food, and the architecture. But these differences are minor, even exciting, and will not take long to get used to. Far harder to deal with are the differences in human interactions and philosophies.

◆ The hardest thing is the big difference between the American and Chinese cultures. My way to adjust is to find my own way to *combine* the two cultures. This is the thing I am still working on right now. (Graduate student, citizen of the People's Republic of China)

When two cultures meet, there are bound to be dissimilarities that evoke annoyance and frustration, as well as admiration and respect. Some of these differences are mundane, but can cause frustration. For instance, foreign students may think Americans are obsessed with being on time and sticking to a schedule: Americans seem to look at their watches very frequently, get upset when someone is late, and rush away from friends to make a scheduled appointment. You may be used to a slower, more relaxed form of existence, and may find this concern over time and schedule bothersome. Likewise, foreign students often comment on how ignorant some Americans are about the rest of the world. They may interpret this as narrow-mindedness or self-centeredness, and it may insult or disappoint them.

♦   Americans don't seem very curious about other cultures, and that bothers me. They can go through life without knowing anything about geography. (Graduate student, citizen of India)

Most importantly, all international students miss their families and friends back home, and they may not receive much help or emotional support from the people around them. They may view Americans as superficial people who have shallow friendships.

♦   The student body within the department did not provide any support, and did not include foreign students in their social engagements. It was through members of other departments, mostly foreigners to begin with, that friendships grew. I never felt I was fully accepted by the faculty as were the other students. A second issue for many students from warm weather countries is the winters. Invariably, the short and cold days of a very long winter are very hard to contend with. (Graduate student, citizen of Mexico)

If students feel a lack of emotional support, part of the problem is graduate school itself—it is not a nurturing environment, and it can be an isolating experience. More importantly, many Americans feel awkward because they do not know if foreign students want to socialize with *them*! It is therefore important that international students participate in department activities and social events. Getting to know other people, Americans and foreigners alike, is easier when they share your hobbies and interests, so think about joining a few clubs or groups (chess, music, sports groups, etc.) or organizations (especially international student organizations). Food is a great attraction, so you may want to get together with other international students and create "international evenings," sharing foods from around the world. Inviting American students and professors to these affairs would be a wonderful gesture, and would engender interest and respect that will surely carry over into the classroom and the lab. Making friends with people from other cultures is one of the great experiences of graduate school—for Americans and foreigners alike.

♦   Nobody disturbs you because you are a foreigner, but nobody accepts you among them either. In general, Americans tend to ignore foreigners, probably because there are many of them. I personally didn't want to carry my culture with me to the U.S.; however, silent rejection of my existence pushed me to stick with my own culture, which caused a conflict in my life. Almost all Americans (especially Southerners) feel that maintaining their normal and private life is much more important than communicating and sharing something new and different with another person. People live so turned into themselves, so alone; as a result, no one trusts anybody.

The main difference between the educational system in the U.S. and Turkey is that whatever I learned in the U.S. was up to date. Researchers in the U.S. have to pick research topics that have a good chance of getting funded by a grant. As a result, they often don't work on their real interests. In Turkey, it is very difficult and expensive to find recent books or literature. Things are quite old, and state-of-the-art learning tools are a luxury. The U.S. learning environment has much more discipline, although, and this is the best part, you are still free to discuss and disagree. (M.D., graduate student, citizen of Turkey)

♦   The learning environment here is very different than in Turkey. It is more systematic. But, we have more dedicated and better teachers in Turkey. The tools needed for education, i.e., computers, laboratories, etc., are very satisfactory here. I like America. Every kind of freedom exists here. However, American culture does not fit into my culture. Most of the people here are too money-oriented. Money is an important factor in life, but it is not the main thing. The main thing is having good relationships with people. One problem here is that competition affects people's relations. They feel that they have to be selfish to do good research. It is hard to find good cooperation. Another thing that I don't like is American food. It is very standard. No variety. Taste is not pleasurable. I am thankful that I learned how to cook. (Graduate student, citizen of Turkey)

There are no official statistics on how many foreign students remain in the U.S. after getting their degrees, although 50 percent has been estimated. Those who want to stay worry a great deal about whether they will get a job (the requirement for remaining in the country). Jobs in science are hard to get, however, for foreigners and Americans alike.

♦ My biggest insecurity is what comes after the Ph.D. It's a big issue for me, because it decides whether I stay in this country or go home. I can legally stay for one year if I find a job that is a continuation of what I am doing. After that, I have to get someone to give me a work visa and that is out of my control. They have to show that I am the only person for the job and that they can't hire an American to do it. (Graduate student, citizen of Poland)

♦ My first difficulty here was English. I almost quit school after my first biochem class. I also had difficulty interacting with American students. For example, what is happy hour?

But I think these things are not big problems once you adjust. The real problem occurs after you finish your Ph.D. Most industries require citizenship for applicants. Academic jobs are no different. You need to be a citizen to apply for some grants. (Graduate student, citizen of Korea)

14

On the Art of Scientific Writing

♦ I was attracted to academia for the stereotypical reasons: I believed science was a field where I could sit in my chair by myself, pump out good work, and be recognized for my achievements. The biggest learning experience for me was in the sphere of communication. I had to *learn* to appreciate that *communication with peers and with the general community*—via posters, meetings, and papers—was, in fact, one of the most important aspects of the scientific career. What did I care if the axes of my graphs were not perfectly labeled? Weren't readers smart enough to figure out what I was saying? Yes, probably so, but they are also busy and could not be expected to pick out whatever gems I had to offer from a morass of discussion. (Ph.D., Physics, Princeton University)

Just *doing* good science is not enough to establish one's reputation as a scientist, nor is it enough to secure employment in the field. Scientists must be able to *communicate* their science, to tell others the content and meaning of their work. Work that is never communicated, via talks or posters at professional meetings, seminars, journal articles, or books, is useless, and may as well have not been done. Not only will such research never get the chance to be critically evaluated by others in the field (one of the hallmarks of the scientific process), but it also will not contribute to the work of others, who can use it as a building block for their own research. Uncommunicated research will also do nothing to enhance your reputation in the eyes of your peers and potential employers. Every scientist, therefore, must be able to write and talk about his work, and graduate school is the time for practicing the skills involved.

The emphasis in this chapter is on writing scientific manuscripts, a task for which grad school offers abundant opportunities to gain experience. Grad students write progress reports on their research, mock grant proposals (for classes/qualifying exams), actual grant proposals (for fellowships/student grants), abstracts (i.e., condensed summaries of work), journal articles, book chapters (possibly), and, of course, the dissertation. They also prepare posters (i.e., printed descriptions of performed experiments pasted on posterboard), for display at professional meetings. All of these various writings share common characteristics.

The journal article is discussed here as the prototypical example of scientific writing. Writing experiments up for publication in scientific journals is a weighty, intimidating prospect, and it is indeed unfortunate that the ability to do good science does not necessarily translate into the ability to write well. In fact, some of the best and most rigorous scientists are justifiably insecure about their writing abilities. Fortunately, one does not have to be a great writer to succeed in this business; one does, however, have to be a careful writer.

Scientific Writing Is in a Class of Its Own

A scientific manuscript is painstakingly constructed. Words are used sparingly, and each word is carefully chosen to be as definitive and also inclusive as possible. Adverbs and adjectives are generally scientific in nature. Sentences are tight and cut to the bone. There is little or no repetition, and little embellishment. Issues that are not germane are excluded. The result is an unusually dense, compact composition in which the typical word is an umbrella under which lie layers of conceptual meaning.

Because *every word counts,* and looseness and imprecision are anathema, scientific manuscripts are typically prepared (and read) slowly and deliberately—one does not rush ahead. They are then revised and rewritten many times before satisfaction is achieved. (Some authors prefer to do the first draft relatively quickly, writing down all their ideas without regard to organization; later they voraciously edit, scrutinizing each word.) Such laborious efforts are the only way to ensure the precision, conciseness, detail, linear organization, and circumscribed structure that is required of this type of writing/publication. In one sense, these strict requirements make the writing task time-consuming and difficult; writing becomes the solving of a long series of "word puzzles." In another sense, the requirements create a rigid framework that can actually make the job easier than if one were "burdened" with the freedom to shape and expound at will.

The only way to learn how to write scientific papers is to read scientific papers—lots of them. You may not be consciously aware that learning is taking place, but after awhile, you will assimilate general aspects of structure and content, and you will develop a pretty fair sense of how a paper flows, what words are commonly used to describe and characterize phenomena in your field, and what types of information are important to include. The content demands of a typical paper will become even more apparent when you actually have to look up some particular journal article in order to get information. When you seek out printed specifics so that you can perform a technique, run an assay, or figure out why you could not replicate the author's results, the necessity for publishing practical details becomes obvious.

A useful training method for developing skills in scientific writing is the use of an "historical" approach. (Try this at an early stage of your research, before you actually need to write a paper.) Say your lab is studying some aspect of phenomenon A. Get a copy of a recent paper that your lab has published, and circle all references that deal with phenomenon A. Get copies of all these papers, and repeat this process for each. With enough repetitions, you will eventually get back to the original papers describing the discovery of the phenomenon. Start your reading with these original papers, and then read the others chronologically—as you work your way up in time, it will be like seeing history in the making. Besides teaching you the historical background of your lab's focus (something you should learn anyhow); how ideas develop, are tested, and are challenged; and how these processes are expressed in the literature, this technique will also make you aware of some basic realities of scientific writing:

1. All papers have a similar basic format. In general, all journals require submitted manuscripts to have sections devoted in some form to abstract, introduction, methods, results, and discussion/conclusions. While, of course, the specific content of any one of these particular sections differs from paper to paper, the general scheme of construction of the section can be pretty similar across papers. This is es-

pecially true for papers in the same field dealing with similar techniques (some of these will take more effort to understand than others, commonly a reflection of the clarity of the writing). Thus, you can get a good idea as to how to approach each section by looking at other papers in your area. By building up your manuscript upon the general "scaffolding" that you observe, using your unique subject matter and data, you can generate a pretty decent first draft of a paper. Note, for example, the following in the papers that you read:

What is the introduction like? How does it start out? How narrow or broad a background is described and referenced? What types of things do the authors assume the reader knows, without the necessity for an explanation or a reference? When and how do they bring up the issue of their present work? Do they say why their work is important?

What is included in the methods section? What techniques or methods did the authors feel comfortable leaving out, glossing over, or referring the reader to another paper for? How detailed were the descriptions of the remaining techniques/methods? Do you notice that authors often use a relatively "stock" format for wording the descriptions of particular techniques?

How are the results described? What types of things go into the text versus a table, a graph, or a diagram? How are statistical analyses worded?

How is the discussion (conclusions) section organized? Do the authors start right off summarizing their data, or do they discuss the background of the work first? How does the discussion section differ from the results section? How much do the authors theorize? How do they fit their results into the "big picture"? How do they address an experiment's shortcomings in acceptable garb? How do they explain away loose ends or inconsistencies in their data? How do they explain results different from those that other researchers have gotten? How do they conclude the paper?

The observation that journal articles in the same area tend to have similar and predictable construction schemes means that when you prepare a manuscript, in a sense, you are not starting from scratch; the inherent common structure of many papers makes the writing of the particular sections somewhat of a fill-in-the-blank task. Sometimes common phrases are repeated in paper after paper. (As a colleague once said: "I think there may be only one original paper in the universe, and nothing new has been written since; all others copy off of it.") This is not immediately obvious, especially to a inexperienced graduate student who can be overwhelmed by the complexity of a typical paper, but it becomes more evident the more papers you read and write. Tackle each section of a paper as a separate challenge, *using papers that you respect and that cover techniques similar to your own as guides for structure and coverage.* You probably want to leave the abstract for last, as it is easier to piece it together after you have all your ideas from the other sections in place.

2. When a lab publishes a series of papers on a particular subject, it is usually only the first paper of the series that is difficult to write. Succeeding papers often follow the same basic plan as the first, and then add on a few new twists. Thus the author of the first paper on a phenomenon or a particular subject has the responsibility of gathering all references relevant to the topic, interpreting the topic in the context of past knowledge, assessing the topic's meaning, and logically arranging all of this in the paper's introduction and discussion. In succeeding papers, the general format often remains the same, and many of these lines of thought are "borrowed." Thus, the first paper sets the stage for all others.

3. Authors rarely make bold declarations. They rarely portray their findings as the final "truth." There is a self-protective "holding back" ever-present in their writings. In essence, what these scientists are whispering to their readers is "we *know* that even the most sensitive instruments can give erroneous readings; that mistakes can creep into our methods; that unknowingly we may select a sample population that is skewed; that *in vitro* effects may not mimic *in vivo* ones; that misinterpretation is possible; that $p < .05$ still leaves room for chance occurrences; that we only get glimpses of reality; that appearances can be deceiving; that a theory is only a theory. We know all these things, and we want *you* to know that we know all these things. If sometime in the future, our data or conclusions prove wrong, let it never be said that we believed so strongly and close-mindedly that we never considered that this could happen." After all, isn't open-mindedness the sign of a true sci-

entist? It is important, therefore, when you write a scientific paper, that you make every effort to appear "wishy-washy." While your data in the results section should be clear and strongly presented ("The muscles of ten of ten animals showed homogeneous populations of fast twitch fibers; there was no significant difference in EMG potentiation before and after treatment [B = 6.0 ± 2.8; A = 3.1 ± 1.9; $p > .05$]"), your wording in the discussion section, in contrast, should be noncommittal and cloaked in qualifiers. It is common therefore to see the following expressions used by modest researchers:

> The data *seem to indicate* . . .
> These results *appear to show* . . .
> The evidence *suggests that* . . .
> This *could mean that* . . .
> Thus, this enzyme *may be* . . .
> Thus, *it is likely that* . . .
> Thus, *it is possible that* . . .
> These data *imply that* . . .
> This nucleus has been *implicated in* . . .
> This *argues strongly for* . . .
> This result was *probably* not due to . . .
> The *apparent* difference between . . .
> The *reported* action of . . .

You will see the phrases "We find no evidence that X affects Y" or "Treatment A results in increased activity levels," more commonly than the bold "X does not affect Y" or "Treatment A causes increased activity levels" (the use of the word "causes" is a particularly sensitive issue, as causality in science is hard to prove). Authors will dance all around the results rather than write anything scientifically incriminating. "These effects *provide support for the hypothesis that* factor K acts on the cell membrane" is much safer than "These effects show that factor K acts on the cell membrane." (With the advent of certain precision technologies, scientists in some fields, such as molecular biology, have become confident enough to indeed unhesitantly make some bold declarations in their papers ["These results show . . ."]. Students should thus observe what is considered conventional in their particular areas.)

4. Authors try not to overinterpret their data. Although they can and should hypothesize what their results mean and

where they fit in the overall scheme of things, they cannot make giant leaps of intuition. They cannot hypothesize too far beyond the next logical step, or most journal reviewers will jump all over them.

5. As we have noted authors often go out of their way to be cloudy and nebulous when it comes to the strength of their arguments. However, they are extremely precise when it comes to their choice of words and the meanings these words carry. There is no messiness or imprecision allowed here. Every word means what it says, no more and no less. For instance, they will write:

"Our results suggest that this transmitter *contributes to* the regulation of the hypothalamus," instead of "Our results suggest that this transmitter *regulates* the hypothalamus" (correctly leaving room for additional regulators).

"We *believe that this is the first report* demonstrating a relationship between potential vorticity and the zonal wind," instead of "*This is the first report* demonstrating a relationship between potential vorticity and the zonal wind" (in case there was a paper that the authors missed in their literature review).

"Our results show that *LHRH-like* immunoreactivity is present in the thymus," instead of "Our results show that *LHRH* is present in the thymus" (in case the labeling is nonspecific).

"This result is *consistent with* a core that is solid," instead of "This result shows that the core *is* solid" (since solidity has not been proven).

6. Many authors also tend to be careful to give due credit to alternative possibilities or alternative explanations. This is sometimes carried to extremes, but they want to show that they have considered all avenues. You therefore may run into statements such as these in the conclusions section of papers:

"Although this is the only area of denitrification that we observed, the possibility exists that some other, as yet unexplored, aquifer in the region is also so affected."

"Another possibility is that some unidentified aspect of the testing arena is responsible for the effect."

"Although unlikely, it is possible that the presence of polycyclic aromatic hydrocarbons results from processes other than biological."

7. Some authors try to ward off other researchers who might consider entering their arena by implying that certain stud-

ies (logical extensions of their described work) are already in progress. Whether or not such studies are actually being currently pursued in their labs, is anyone's guess (sometimes they are only being thought about). Areas of investigation are unrestricted, however, and any other scientist has the freedom to "jump in" and pursue research on the topics (although some may be afraid of "stepping on toes"). Attempting to gain priority on future work by implying or stating that there is a work in progress, is not uncommon; it is up to the individual to decide if this is something he or she wants to do.

Authors will write something like:

"We are currently exploring the possibility that. . . ."

"Whether X results in Y remains to be determined." ("Remains to be determined" or "is not known at this time" is also useful for covering up an embarrassing lack of experimentation on certain obvious questions whose answers would make the study more complete. These phrases indicate that the questions, at least, have been noted as relevant.)

8. The assignment of authorship on a paper can be confusing, and the strategy employed is often field-dependent. In biology, the first author listed is supposedly the one who wrote the first draft of the paper and did most of the actual work. This is often a student or postdoc. The last author listed is usually the head of the research group (i.e., the professor whose grant paid for the research and perhaps in whose lab the research was done); often the work was his idea. Middle authors are those who assisted significantly with the work, but did not perform the bulk of it. This order is sometimes altered. A lab head may prefer to have his or her name listed first, and then the person who did most of the work is listed second. As you may imagine, since first and last authorship carries the most prestige, the order of authorship can be a point of major disagreement among researchers on a project.

Writing a journal article or any scientific piece is a test of deep thinking and reasoning. Similarly to doing the experiments themselves, you must be obsessive enough to consider everything of relevance. Your subject must be carefully dissected, you must have a logical flow of ideas and arguments, you should discuss the meaning of your work, you cannot hypothesize or conclude anything that is not indicated by the data (have minimal speculation), your wording must be precise, your organization should be excellent, and your sentence transitions should be smooth. A journal article is a reflection of yourself, and it is the way that others in the field get to know you and form an opinion of you. Interesting and thoughtful papers will get attention and win respect; it also helps to write enthusiastically, persuasively, and well.

Luckily, as a student, you will have experienced coauthors on your papers (including, but not limited to, your advisor), who will help you with the writing. They are likely to send your early drafts back to you strewn with comments in glaring red ink about corrections to be made (be sure you understand why the change was necessary). Virtual or real papers will flow back and forth between you and your coauthors many times before they are considered worthy of sending out to the journals, and each time will be a learning experience for you. Eventually, you will develop a feel for the task. Nevertheless, if you are like many published scientists, journal-article writing may remain a frustrating, although doable, challenge until the end of your publishing days.

Which journal a paper is submitted to depends on the area of the research and the importance of the findings. Most journals (*Blood, Journal of Immunology, Brain Research*, etc.) are specialized, and only accept papers in circumscribed fields; within a field, some journals are more selective than others. The prestigious journals *Science* and *Nature* generally accept papers from most scientific fields, but they are extremely demanding, and manuscripts sent to them should contain unusually interesting and exciting findings that promise a reverberating impact. Students should expect that journal reviewers are going to find fault with their manuscripts and ask for revisions (perhaps even additional experiments) before they agree to accept them for publication. It is the very rare manuscript that is accepted outright during the peer-review process. Students should also know that if they do not think the reviewers' comments are justified, they can write back and politely explain their position, or they can submit their work to another journal. Papers that are turned down for publication because they are fundamentally flawed in experimental design should not be submitted again.

♦ I'm still learning how to write papers—I've only written one. I pick the journal that I want it to go into after consultation with my advisor. I read the "author's instructions" that is available for every journal. My advisor gives me feedback—I write something, hand it in to her, she reads it, and hands it back with comments. Sometimes I disagree and keep the part as it was—often I don't. (M.D./Ph.D. student, department withheld upon request, Brown University)

There are a number of good books out on technical and practical aspects of writing scientific papers. Some of these are noted in the References and Additional Reading section of this book.

15

What Should Your Goals Be While in Graduate School?

It used to be that, during science's heyday, one could get by largely on intelligence and a desire for knowledge alone. If you loved worms and wanted to do nothing but watch and study them, somewhere in academia there was probably a place for you. Those days are no more.

♦ The overwhelming feelings I have at this point in my life, having just finished my Ph.D. and begun work in a postdoctoral position, are worry and disappointment. I worry incessantly about my chances for landing a job doing what I was trained to do. This is not because I feel incapable. I am just continually bombarded with stories of failure and negativism from those who have gone before me. Reports in various scientific/academic journals are likewise bleak. Now that university/governmental purse strings are tight, the age-old recipe for success in science comes up lacking, and those who have invested such a large portion of their lives in this endeavor are left, after all of their sweat and tears, with little or nothing. The large majority of the new generation of scientists must now face the grim reality.

I can't help but feel that I have toiled away a large portion of my life in poverty, all to obtain a degree that no longer opens doors. While my friends who barely graduated from high school are paying on their twin-garage house, their two cars, and the boat, I am left with loan debts and bleak job prospects. Though people do not go into science to make a lot of money, it is nonetheless insulting, after all of the work I have done to get where I am, that I am paid $19,800 as a postdoc and receive less in the way of health insurance than does an individual on the secretarial staff. Though I am disturbed that I have come this far and still need to worry about either having to go on "the postdoc treadmill" or cut my losses and run, I try hard not to dwell on these possibilities. I have invested far too much to turn back now. (Ph.D., university/department withheld upon request)

Today, the idealistic but unrealistic young scientist is in for quite a shock. There are not enough faculty positions in academia to accommodate the pool of Ph.D.s looking for jobs, and funding for many areas of research is limited. Job searches take longer than ever before, and it is estimated that in many fields there are at least 300 postdocs applying for each available faculty position. Many researchers take two or more successive, postdoctoral positions (a phenomenon known as the "postdoc holding pattern"), or settle for part-time or temporary employment—all in hopes of *eventually* obtaining a tenure-track faculty position. Many others are underemployed, taking positions that do not require a Ph.D. Even those who are eventually hired by universities are facing troubles down the road—the percent of federal grant applications that are funded has reached record low levels.

The present competition for faculty positions and research money has made for a disturbing academic climate. Scientists are now obsessed with appearances and productivity. Trivial but manageable scientific riddles can now look more appealing than meaningful but unwieldy ones. Techniques have been exploited for all they are worth, and hands sometimes seem more useful than brains. Fearful people are publishing like crazy, but the quality or value of some of their papers is debatable. Allowing ideas to leisurely simmer and unhurriedly coalesce is a luxury associated with less hectic times. While idealistic attitudes and quality research persist in many labs, the negative trends are prevalent enough to cause alarm. As this professor, who received his degree from a prestigious research university, reveals:

◆ Graduate school ain't what it used to be. I was in a laboratory in which the intellectual framework spanned the two hundred years of the subject's existence, and the philosophical framework, two thousand years. My program focused on intellectual thinking. We grad students may not have ranked up there with Leibniz and Descartes, but we tried; they were our scale of measurement. Because of the special nature of our advisor, we students did not obtain much practical experience obtaining money for our science or writing papers. However, we could bullshit with the best of our peers, and we were usually more interesting, and a tad more original.

The current graduate program seems to have reversed the important objectives. The most important goal seems to be the job market. The intellectual goal has been minimized, and success is publishing papers. Theses are three published papers, tied together with an introduction and a summary. A thesis is one of the few opportunities to think and speculate; the current form of the thesis minimizes the intellectual goal, and focuses on marketability. My advice: if a graduate student does not strive for understanding beyond the experimental basis, he/she should resign, obtain a master's degree and become a great lab tech. Science is fun, but the stress from the current job market and funding levels in nonmedically oriented science is no way to live. (Ph.D., university/department withheld upon request)

Another professor writes:

♦   When I was young, I had the idea that science was a contemplative, peaceful, and rewarding activity where people were civilized in their behavior and reasonable in their attitudes towards each other. As time went by, and as I learned about how science is really done, especially now, I would describe much of it as something more akin to a money-grubbing rat race. (Ph.D., Biology, Texas A&M University)

Many idealistic students, too, are disturbed by the poor job market, and the practical, meat-and-potatoes academic atmosphere found in some departments:

♦   I had a lot of misconceptions concerning graduate school. I thought that: there would be a lot of discussions about science and related topics; it would be expected that you would do a lot of independent thinking; a rigorous academic education would be insisted upon; your Ph.D. would mean something. (Ph.D., Neuroscience, university withheld upon request)

It is not a pretty scene out there in the marketplace, and students are forced to grow up too soon. In order to have a chance at the limited opportunities around today, young scientists are forced to start working on their careers *while still in school.* If you want to have a position in academic science, if you want to compete successfully against your determined and desperate peers, you have to have a healthy amount of self-initiative and you have to have a *plan* in mind from early on. You must start "playing the graduate school game" as soon as you arrive at the department. Having such a premeditated strategy, as ugly as that concept may be, is a practical necessity, a survival tactic. It is the reality of the present situation.

Goals to Accomplish

♦ Faculty positions are *very* hard to obtain. Still, they do exist, and people are getting them. If academic research is what you really want, I say: Go For It!! (Ph.D., Neural Science, Indiana University)

Somebody is getting each one of those faculty positions regularly advertised in the back of *Science*. Competition is stiff, but there is no reason why, if you step out onto the "playing field," you cannot have a good fighting chance at one of those slots. Described below are the ultimate goals of the "graduate school game." These are goals that you should seek to accomplish *while attending* graduate school. Many have been discussed in other chapters but are listed here for easy review. If you want to be eligible for those precious faculty positions out there, keep these objectives always in mind. If you wait for trial and error to teach you, it will be too late.

Form a Networking Framework

Help the scientific community get to know you. Meeting and interacting with researchers at your own and other universities, gives you a chance to naturally manifest your sensibility, maturity, and intelligence.

♦ One of the most important parts of getting a job is letters of recommendation. Although one has little control over this, it is important to develop relationships with faculty other than your advisor. These can be people on your thesis committee, collaborators at your institution, or (most impressive) people at other institutions. You will need letters of recommendation when applying for postdoctoral positions, postdoctoral grants, and faculty positions, and it is helpful to have a number of people who can write you letters. (Ph.D., Genetics, Yale University)

Become Proficient in Some Useful Techniques or Mathematical Skills

People who can bring in new techniques or skills, especially those that are considered "hot," are coveted. Your competence, and presence on the cutting edge of science, will make you marketable for both postdoctoral and faculty positions. If you cannot achieve such expertise as a student, aim for it as a postdoc. Do, however, become competent with computers while in school; all scientists must be facile in this area.

◆ Things seem very different now than they were when I was in graduate school. I never really worried about whether I would be able to get a job later on or whether I was learning the things that would make me employable. I decided early on that I wanted an academic position, and there was an unwritten sense that this was the expected path for the best students; only those who were in some sense "not good enough" took positions in industry or government.

My advice to students now would be to work really hard (which I did) and get the broadest possible experience (which I didn't do). Don't do just one thing, however well, for four or five years. Learn as many techniques as possible, work on many different systems, and consult with many different professors in your department, so that you'll have more than one person who knows your work well enough to write you a meaningful letter of recommendation. Also, read the literature, attend seminars by outside speakers, and take all opportunities to go to scientific meetings and find out what's happening outside your institution. (Ph.D., Chemistry, University of California, Berkeley)

Make Sure Your Research Tells a "Story"

When it comes time for you to give a seminar as part of an interview for a postdoctoral position, you do not want to portray your dissertation experiments as isolated phenomena occurring in a vacuum. Hopefully, you picked research projects that were addressing *important* issues, issues brought to light by the works of others. Your own work presumably clarified these issues, and probably opened up some *new* ones, which you explored with still more experiments. Together, your experimental results provide a more detailed and cohesive view of reality than was available before you began working. So, when you discuss your research, you should have an interesting and flowing "story" to tell. This should include the past state of knowledge, how your findings altered or added to this former view of reality, the additional questions that your results brought up, the results of the additional experiments you performed to answer these questions, and the implications of all your work in terms of broader issues. Do not get lost in the *details* of your experiments; it is more important to relate your findings to the "bigger picture." If you isolated and characterized a new gene, how do your findings relate to more comprehensive topics, such as gene regulation/expression? If you reported a new molecular structure, what do structure/function relationships imply about its function? If your work revealed trial-and-error learning of predatory behavior in spiders, what does this say about the supposedly rigid, instinctive behavior of such animals?

Learn to Write a Grant Proposal

You may already have some experience with this task. Perhaps you applied for a fellowship and had to write a research proposal; perhaps you had to write a mock grant proposal for a course, or as part of your qualifying exam. If you do not have experience in this area, get some soon. It is very possible that you will be asked to apply for a grant as a postdoc, and you certainly will be as a faculty member; even industrial scientists write proposals for federal grants. You can wait and learn as a postdoc, but it is better if you get at least a start now. Ask your advisor if you can read some of his or her (funded) grant proposals. Study them carefully and note their organization, wording, and logic. Remember that the three most important characteristics of a good candidate for a faculty position are (1) Ph.D./postdoctoral work performed at quality institutions, (2) good publication record, and (3) fundability.

♦ I volunteered to help with my advisor's grant proposals because I wanted to learn how proposals are written. We used to sit late at night in the lab and I typed the grant for him. From this experience, I learned how he approached a problem, and this helped me to develop the art of scientific writing. (Ph.D., Physiology, Kansas State University School of Veterinary Sciences)

Publish Two to Three First-Authored Papers

Getting publications is a crucially important goal of graduate students, postdocs, and more advanced researchers. If your work is not in print, there is really little way for the world to recognize what you have done or what you are capable of doing. Without publications, your chance of getting a respectable postdoctoral position is diminished and, thus, your likelihood of success further down the road is limited. Although you may have accumulated an enormous amount of knowledge during your graduate years, and although you may have been taught to think creatively, carefully, and deeply, you will have little means of proving this to future employers except by means of your publications.

> ◆ It may not be such a dumb thing
> To adhere to this short rule of thumb thing:
> In attempting to steer
> Your way through a career,
> Don't try to BE someone, try to DO something!
> (Ph.D., Geophysics, Columbia University)

Become a Good Public Speaker

Public speaking is a key tool of the trade. Being able to give good talks that are not cluttered with detail, that are easy to understand, and that make your work sound interesting will enhance your reputation significantly and increase the impact of your research. To perfect your skills, seek out every chance to speak publicly that you can, and ask acquaintances in the audience for honest feedback. Eventually, you will want to broaden your exposure by presenting your research at professional meetings. Here, with the aid of slides, you will talk for ten-twenty minutes (followed by a five-ten-minute question-and-answer period), possibly in front of a large audience. Fear will be unavoidable, but don't let it stop you.

> ◆ I was terrified of the seminar I had to give for the whole department during my second year, but after extensive grilling in practice talks for the group, the actual seminar went very well and really helped my confidence. (Ph.D., department withheld upon request, University of California, Berkeley)
>
> ◆ I don't remember any of my first talks. I was so nervous, I couldn't think as I said things. I didn't have any of the skills

needed to interact with the audience. Once you go through the fire though, you learn that it is not so bad, and that people are not there to crucify. I also reached the point where I am comfortable saying that I don't know something. My lab listens to my talks beforehand, and gives advice—what to take out, what to focus on, body language, etc. (Graduate student, Neurobiology and Behavior, Columbia University)

♦ When you give a presentation, or when you write a paper or a grant proposal, never tell all you know or are thinking on the topic. Otherwise, you will have nothing to say if questions arise. Leave some things purposely open-ended so that certain questions, for which you know the answers, are naturally brought up. People love to have something to criticize or ask about, and this way you will come across as pretty sharp. (Ph.D., Psychology, biopsychology research emphasis, Vanderbilt University)

♦ In general, if someone asks you to give a talk, say YES! If you have the opportunity to volunteer to give a talk, do so! Giving a talk forces you to take a hard look at your data and what it means, and gives you crucial practice at communicating your ideas. These skills are critical for being a successful scientist. (Ph.D., department withheld upon request, Princeton University)

Finish Graduate School within a Reasonable Amount of Time

Students who cannot complete their graduate programs within an acceptable time frame (which varies from department to department) may be viewed as procrastinators, or worse. Obviously, this is not how you want to appear. However, in a situation like graduate school, where one is allowed to proceed at his or her own pace, procrastination is a very easy state to fall into and many a student ends up getting "stuck" at one stage or another. It takes determination not to procrastinate, to keep busy and make steady progress. You have to be your own manager, and set deadlines for yourself. Get a calendar and schedule when you want to finish your courses, start experiment X, begin writing the dissertation, etc., remembering that most things will take longer to do than you expect. Also write down important dates as they come up—when you will be presenting at a meeting, when you are signed up to use the darkroom. Make all efforts to stick to your schedule. Handle your duties as if you were working at a job: work even when you do not feel like it, make use of blank time (yet remember that

chatting with your peers is *not* a waste of time!), and make constant progress. Efficiency is key; it is all a balancing act.

♦   I strove to be efficient. While my gel was drying, I used the
time to mix up solutions. While the electrophoresis was running,
I would graph my last data set, or study for an upcoming exam. I
usually managed to finish my workday by 5 p.m. And I seemed to
get a lot more done than those people who wasted their hours and
worked until midnight. (Ph.D., Biology, California Institute of
Technology)

♦   Learning to be more organized and less of a procrastinator is
the single most valuable thing I learned in graduate school. My advisor is very organized and punctual, and I learned a lot about time
management just through my interactions with him. I learned to
finish things well ahead of their deadlines. This strategy applies to
writing abstracts, making posters, preparing talks, etc. Being
ahead of schedule not only prevents anxiety, but it allows time for
editing and revision. (Ph.D., Genetics, Yale University)

♦   Grad students have to get organized on their own. *They can't
rely on their advisor for this.* For one thing, everyone has to have
a calendar to write down when a talk is happening, when an abstract is due, etc. I plan when I am going to take the qualifying
exam and other hurdles. Because there is no set structure in grad
school after your classwork is finished, *you have to make some
artificial structure* and organize ways to reach your goals. (M.D./
Ph.D. graduate student, Neurobiology and Anatomy, University
of Rochester School of Medicine)

Develop Street Smarts

Your graduate years are also a good time to observe how the professionals play the game. Keep your eyes open, and watch how the
faculty think and conduct themselves. Forward-moving professors
have many contacts whose studies they keep up with. They obses-
sively watch out for new, hot areas of research that they can move
into, and for fresh, cutting-edge techniques that promise to yield a
new dimension of information; as a result, their labs are never static, but flow and grow with the times. They keep on top of the
funding situation and gear their research accordingly. Tough and
determined, these professors stand up for themselves with respect
to lab space, salary, and other issues. They interact frequently
with their peers, and make sure that they learn the ropes and are

passed down trade secrets. They are self-promoters and are politically active, serving on professional committees and holding office. Effective faculty keep themselves organized, are aware of what is going on in their labs, stay within their budgets, manage and motivate their lab members, and handle lab conflicts. They are entrepreneurs, businesspeople, politicians, managers, and psychologists—without having taken a single, pertinent course.

◆ If you want to have a career in academia, you should know how to network and work with other people. Networking involves getting feedback from your peers—science works better when minds come together to solve a problem. My advisor knows everybody in his field, and generally what they are working on. His lab survives to some extent because he incorporates their ideas (suggestions) into his lab, and he collaborates with many people. I suspect that this form of networking is a very important thing, and I am fortunate to see how someone does it. These are survival issues. (Graduate student, Ecology, Evolution and Behavior, University of Minnesota)

Secure a Postdoctoral Position in a Well-Respected Lab

Postdoctoral training wasn't always a prerequisite for obtaining an academic job, but it is today. This poorly paid research position, supported by a professor's grant, an institutional training grant, or an individual fellowship, usually lasts two to three years. Just as it is important for a graduate student to work for the best possible person he or she can, so it is (certainly even *more* so) for a postdoc, whose reputation will reflect the reputation of the postdoctoral sponsor, and whose career can be helped by the connections and influence of that sponsor. The focus of the postdoctoral research is also important (fields that are likely to have a long funding future are the most practical, of course), as people tend to make their postdoctoral labs the bases of their own labs.

Check out postdoctoral sponsors as you did advisors: find scientists whose work you find fascinating, note if they publish in the best journals, and speak to people who have worked for them. Pay attention to where their former postdocs have ended up; whereas a major role of graduate student advisors is to get their students postdoctoral positions, the role of postdoctoral sponsors is to get their postdocs faculty positions (or good nonacademic positions). Start thinking about whom you want to work for at least two years in advance; it can take that long to establish a relationship with a sponsor and to wait until there is an opening in the lab. An alternative route, recommended only if you have no sponsor in mind, is to wait until you know approximately when you will be finished, and then check to see who has a position that will be available. You can learn about postdoctoral openings (and other employment opportunities for scientists, including faculty or industrial positions) from acquaintances, notices sent to the department, placement services at professional meetings, and advertisements in professional journals. Do not expect to find postdoc or other academic openings listed in newspaper ads.

Many scientific and academic publications post their classified sections on the internet, and one does not necessarily have to be a subscriber to gain access. Three of the most popular such sites are *Science* (http://www.recruit.sciencemag.org), *Nature* (http://www.nature.com), and *The Chronicle of Higher Education* (http://www.chronicle.merit.edu). Be advised that there are literally hundreds of web sites where job postings can be found. A good place to start your search is at the internet address of the main journal or newsletter in your research area (found on the inside cover of the journal's paper version). There are also internet sites that link job advertisements from several sources. One excellent such site for biologists and other scientists is Employment Links for the Biomedical Scientist (http://www.his.com/~graeme/employ.html); this web site provides links to all the major scientific journals, biotechnology firms, science discussion groups, electronic resume posting services, international job openings, and on-line essays about science. Do not neglect your heart in all of this— it is important that you find work that is *satisfying* to you. If you end up using hot techniques in the lab of a Nobel Prize winner at the number one institution in the country and hate every minute of it, what is the point?

A potential postdoctoral sponsor should be sent a copy of your cv, and a cover letter expressing your interest in the position. Be sure to tailor the letter to the position, emphasizing your appropriate interests and technical capabilities. Mention specific papers that the professor has published, and explain why you found them exciting. If the sponsor is interested, you will be invited for an interview, and may be asked to give a talk on your doctoral research. It is important that you read up on the work of the group that you are visiting, so that you can ask some intelligent questions, and can explain how you can contribute to the lab's research. If you end the interview convinced that this is the group for you, send a note expressing your continued interest in the position.

Many researchers recommend that you do your postdoctoral research at a university other than the one from which you got your Ph.D. degree, so that you are exposed to new ideas, techniques, and philosophies. Some students take a postdoctoral position outside of academia—in a government lab, or in industry—and then return to academia to accept a faculty position.

> ◆ Staying on for a year as a postdoc in the same group was an easy mistake to make, but I lost the opportunity to learn about another research area and observe a different style of management. For example, I would have benefited greatly from some experience writing proposals. (Ph.D., Chemistry, Massachusetts Institute of Technology)

The impression you make as a postdoc can make or break your career. Your most critical goal is to publish, so make *sure* that you will be working on projects that can be successfully completed within the two-to-three-year period. When you are finished, you want to be able to call yourself an expert in your research area, and you want to have projects in mind that you can work on if you are offered a faculty position. Some people take successive postdoctoral positions in *different* areas so that they can portray themselves as experts in more than one field, thereby increasing their chances of matching some department's needs. We recommend that you read *A Ph.D. Is Not Enough: A Guide to Survival in Science,* by Peter J. Feibelman for insightful information on how to present yourself at the postdoctoral interview, and how to make the most of your postdoctoral years.

> ◆ I got my Ph.D. and did a postdoc (University of North Carolina) and then began to learn how "all of this" really worked. There were about 300 applications per entry level (assistant professor) tenure-track slot. The cvs that rose to the top tended to have certain "obvious" characteristics:
>
> 1. Ivy League, big 10, or big 20, Ph.D. institution
> 2. "Exotic" postdoc institution, e.g., Rockefeller, Max Planck, Cold Spring Harbor, in addition to the usual

3. Two-three publications per year (yes, some in *Science, Nature, Cell, PNAS*, etc.)
4. An existing, transferable grant
5. A specialization that fits in with the specialization that was advertised
6. A specialization that could be integrated with that of some existing faculty members, so that the applicant could become a coinvestigator on future grant proposals

In other words, you have a window of opportunity and if you miss it, then you are messed up for life.

There is an implicit fork in the road for you to take sometime in the future: extrapolate from your knowledge of your progress, and either conclude that you have a competitive chance at success, *or* conclude that you will be, or already are, in the "bottom half" of the pool of people in your class, and get out of the pipeline now and into nonacademic pursuits. (Ph.D., Biology, Texas A&M University)

The Professors Speak Out

We thought that we would end this chapter with quotes from a number of professors, each of whom was asked, "What makes a good graduate student?" Perhaps the answers will enlighten you as to what the expectations of faculty are. Maybe they will help you to formulate an image that you can model yourself after. Either way, we found these answers interesting, and think you will, too.

Dr. Floyd Bloom, Professor, The Scripps Research Institute, La Jolla, CA (and Editor-in-Chief, *Science*):

What I look for in a colleague is someone with enough independence to question assumptions that I don't express clearly, and someone with enough energy to carry out experiments that we both regard as "the next crucial steps," and then move ahead when the results so warrant.

Dr. Nancy Forger, Associate Professor, University of Massachusetts, Amherst:

A good student works hard and consistently, is *self*-motivated, does not procrastinate, eagerly reads the scientific literature, and willingly assists others in the lab.

Dr. Mary Harrington, Associate Professor, Smith College, Northampton, MA; and Associate Faculty, University of Massachusetts, Amherst:

Hardworking, curious, independent, responsible, intelligent, lucky, good with their hands. Independent in *thought*—questions what is taught, argues, stands up for their ideas, has some ideas different from "dogma."

Dr. Zev Rymer, Professor, Northwestern University, Evanston, IL:

Good graduate students have the following characteristics:

1. Above all, they must have a clear research goal, and a strong desire to achieve it, against any odds. These attributes can readily offset limitations in intellect and previous training. A student who "does not know what he or she wants to do," or who states that "it all sounds very interesting, but I can't make up my mind," is less likely to succeed.

2. They must interact and communicate well with others. A student can be very smart, but if he or she is not a good communicator, he or she will not succeed in a very competitive job market. This relates not only to teaching, but also to scientific presentations, and to committee and faculty work.

Dr. Terry Sejnowski, Professor, The Scripps Research Institute, La Jolla, CA:

What most catches my attention in a class is a student who asks a question that reveals a fresh dimension to the topic under discussion. Undergraduates are the most likely to ask such a question, perhaps because they have had the least experience with the subject. Sometimes these questions can lead to useful insights and new research directions.

Dr. Ei Terasawa, Professor, University of Wisconsin, Madison:

I believe that there are three fundamental requirements [for a good student]. First, a student should be bright, intelligent, and inquisitive. Since there is an enormous background to understand in any given scientific field and new key scientific discoveries appear every day in numerous scientific publications, high intelligence is an essential for a good student in science. A bright student is also inquisitive and he or she will keep asking, "Why does this happen?" and "What is the mechanism of this phenomenon?" Second, a student should have a creative mind. Creativity is synonymous with the ability to solve problems and it leads to important discoveries. Third, a student should be self-motivated. Hard work is also important in a scientific career, but students will not work hard unless they are self-motivated.

I thought I should add one more paragraph to describe what my professor in Japan used to tell me. (His name was Masazumi Kawakami; he died in 1982.) He said that, to be a successful scientist, a student needs 1) un, 2) don, and 3) kon. These Japanese words translate in English to (1) luck, (2) dullness, and (3) persistence. Here, he assumed that, besides basic intelligence, anyone who wishes to be a scientist will (1) have a certain amount of luck in obtaining a good opportunity (such as choice of professors, jobs, etc.); (2) not be too smart or act so quickly as to overlook the best possibility; and (3) have enough persistence to achieve the goal.

16

Times They Are A-Changing

The situation in science nowadays, as outlined in the previous chapter, is very different than it was even a relatively short time ago. The cover of a 1995 issue of *Science* (vol. 270), depicting a tasseled and gowned, newly minted Ph.D. searching through the newspaper employment ads, tells it all: the academic job market is poor. The causes are many: the race for scientific superiority slowed down with the end of the Cold War, the federal deficit justified a reduction in research funding, the raised mandatory retirement age resulted in a sluggish faculty turnover, and state universities had major cuts made in their budgets. Together, these events have led to faculty layoffs and hiring freezes. Yet the demand for graduate students to teach undergraduates and perform the faculty's research persists: the number of universities granting Ph.D.s have increased, and an over-abundant number of Ph.D.s are being produced.

All fields of science seem to have been hit hard. E-mail bulletin boards are filled with the angry complaints of students and postdocs who are realizing that their chances of getting a job in academia are paltry. Only 31 percent of Ph.D.s who received their degrees between 1983 and 1986 were in tenure-track positions or attained tenure by 1991, and the situation is the same or worse today. Some tenured faculty contend that scientists are no different from other workers, and have no right to feel that they should be guaranteed a job. But surely something is wrong when preparation for one's career is so long and hard, while the chance of actually pursuing that career is so dramatically slim.

♦ Surely a Ph.D. from that well known Midwest state university, the one whose profs were journal editors etc., would get me where I wanted to go. But as I approached graduation, at the age of 27+, I realized I was not competitive in any job market. . . . Five years of equally wasted time on postdocs before I found my present position. As career preparation, grad school was almost worthless.

My advice: unless you are learning something that has clear commercial value and it is obvious that you will be employable when done, don't do it. Now none of this is to say that I did not have a very good time in grad school. I had a blast. I liked my profs. I liked my teaching duties. I enjoyed my sheltered and limited on-campus life-style. My gripe is that the program was economically valueless, yet the profs acted as if it was a big deal. Perhaps they did not know better. Still, universities have continued to crank out unemployable Ph.D.s. I see this every day in my work: we routinely hire [former] postdocs at salaries that are obtainable by anyone willing to work hard who has had perhaps one year of schooling beyond high school. I know this all sounds awful, and terribly negative, but I would not recommend grad school in any of the basic sciences to my worst enemy. (Ph.D., Bacteriology, Iowa State University)

With more Ph.D.s being produced than there are academic jobs, a gap is developing between the secure, older, tenured faculty, and the bitter, resentful, younger scientists in part-time or nontenured positions, who are trying to deal with this new world. However, the world is big, and there are *nontraditional* opportunities out there for adventuresome Ph.D.s. In fact, the National Science Foundation reported in 1996 that *30 percent of Ph.D. scientists and engineers have jobs that are outside of science.* Only 41 percent of science and engineering Ph.D.s are primarily engaged in university research; only 22 percent are primarily engaged in university/college teaching. Furthermore, *more than half the Ph.D.s in the physical and computer sciences work in industry.* Obviously, there are opportunities for scientists outside of the traditional, academic realm (although even some of these, such as certain types of industrial research, are reported to be approaching the saturation point).

In spite of all this, most universities persist in blindly training students for careers exclusively in traditional academic research, careers that may never materialize.

◆   The current system will inevitably change. A system which prepares a large number of students primarily for almost nonexistent jobs simply must change. (Ph.D., Biochemistry, Molecular and Cell Biology, Cornell University)

Indeed, professional journals and newsletters are rich with editorials and articles suggesting modifications in education that may alleviate the problem. Some propose a restructuring of the Ph.D. program, others propound a change in admission policies. The most commonly mentioned suggestions are listed below. Do not be surprised if you see some of them put into practice soon.

1. "Birth control." Some departments have already started to cut down on the numbers of grad students they are accepting, and others are talking about doing so. While many professors are convinced that reducing the size of entering classes is a solution to the employment problem, some are adamantly against this practice. Dr. George A. Kimmich, Professor of Biochemistry and Biophysics at the University of Rochester, explains his personal position on the issue below.

♦ It certainly is true that it is unrealistic to expect as high a percentage of Ph.D. graduates to find careers in academia as in the past, yet should we be taking fewer students into our Ph.D. programs? In my mind, the answer is a loud and unwavering, No! I have a hard time believing that limiting educational opportunities is the way to solve this problem. These are troublesome and uncertain times in science, but the trouble and uncertainty is not limited to that field. Restructuring and making the best use of limited resources is a fact of life in every setting.

Are you happy in your work [as a grad student]? If your answer is "no," or you are indecisive, this can be a danger signal that deserves being heeded. If your answer is "yes," I have no doubt that we have done the right thing in offering you an opportunity for developing your research skills, and that you are doing the right thing in seizing that opportunity. I always advised my daughters that garnering all the education they could muster was absolutely in their best future interests. They might not be able to discern exactly how a given part of their education would be valuable at the time of doing the learning, but I tried to assure them that in one way or another it would indeed be valuable. I believe in this philosophy unabashedly and wholeheartedly and have yet to see exceptions to the generality. *Not being useful in the way originally intended is not the same as not being useful.*

It may not be easy to predict your career path so keep your options as wide as possible by having broader interests than just those related to your thesis project. The world always has room for talent.

2. Shut-downs. Some inferior or (even) small Ph.D.-granting departments may soon be shut down altogether.
3. Revising master's degree programs. Whereas the training of master's students is usually not considered a high priority, the new thinking is that the efforts of these students to prepare for jobs in science that do not require a Ph.D. should be taken seriously. Scholars have proposed the creation of prestigious, high-quality master's programs—some that include courses in business as well as science—to prepare people for positions in academia, industry, and government. MIT has recently created such an elite program in geology.
4. Interdisciplinary training. The crossing of traditional boundaries between disciplines is going to open up new fields, lead to major new findings, and create new jobs. A new emphasis on interdisciplinary courses and research, such as studies that combine physics and material science, or neuroscience, physics and computer science, will create flexible scientists who can work on real-world problems.

    Interdisciplinary training may be especially beneficial to those heading for a career in industrial research or even nonresearch areas. It is a common complaint of industrial employers of scientists that the typical Ph.D. is too specialized, that more versatility and flexibility are needed. For instance, a biotechnical company's project on polymers to assist nerve regeneration might require the research of neuroscientists, mechanical engineers, physicians, biochemists, and chemists, and these specialists must be able to communicate, collaborate, and understand what each of the others can contribute to the problem. In addition to their specialized research work then, the new generation of graduate students should be encouraged to become more broadly educated in a number of fields (by more coursework, if not by research); they also need more *basic* training, in addition to their specialized training, in their own field. This training will increase versatility and adaptability, creating workers that can do a number of tasks at their jobs, and can move easily from one project to the next (a graduate program with such an emphasis is already in place at Northwestern's department of biochemistry).

5. Better mentoring. Academic advisors are often out of touch with what is going on in the Ph.D. job market, and this needs to change. Students must be made aware that a good percent of them will not be able to find employment in the field for which they are being trained. They should be counseled that they can still make major contributions to society by using their scientific knowledge in challenging but nontraditional fields, such as industry, law, politics, secondary-level education, science writing, business, health care, environmental conservation, quality control, urban planning, or government. (It has even been proposed that a variety of types of Ph.D. degrees be offered, each with different requirements, to prepare students for academic or nonacademic employment; many fear, however, that this would water down the Ph.D.) A career in an unconventional sphere, once considered a stigma associated with the poorer students, should today be recognized as a viable choice of a competent professional seeking to achieve great personal satisfaction. After all, the capacity of scientists to be organized and self-motivated, to think deeply and critically, and to solve problems creatively and independently can be put to use in a lot of careers besides research. In fact, Ph.D.s working in government, business, and administration have been shown to have greater job satisfaction than those who work at colleges and universities (Solmon, et al., 1981). Advisors should be able to educate students on the opportunities available, and on the courses or other preparatory activities of which they should take advantage.

6. Collaboration between academic and nonacademic institutions. To ease the transition of scientists planning to do research in a nonacademic setting, it has been suggested that students be allowed to do internships in, or dissertation work in collaboration with, industry or government. A pilot program has already been established around this suggestion. For its part, industry would send scientists to teach at universities, and would help fund graduate education and research.

7. Reduction of time to get the degree. Reducing the number of years required to earn the Ph.D. would allow new scientists to get into the work force earlier (scientists now do not get their first real job until their early thirties or older). Students should be aware that taking a postdoctoral position is not a necessary prerequisite for many nonacademic jobs.

8. More emphasis on teaching. Rewarding and promoting good TAing would aid students who end up at schools where teaching is emphasized, or in industry, where it is necessary to explain scientific issues to nonspecialists.

9. Funding more students by fellowships/traineeships instead of by research assistantships. Research assistantships force students to specialize early, narrowing their education. Students may also be exploited and kept on longer than necessary if they are dependent on their advisor for financial support.

10. Things are fine as they are. Despite the cries of many for change, still others in the profession say it is best to leave Ph.D. programs untouched; after one goes through a program and becomes competent in the critical skills of dissecting and solving problems, they say, he can always pick up other types of expertise (such as business skills) later on the job.

Students Can Also Help Themselves

While a few of the above proposals are already being carried out at some institutions, most are just suggestions on paper. How, or if, the disheartening situation of today's Ph.D. students will be resolved is anybody's guess. As such, what can students do to help themselves? Most students don't think about issues of employment until they are almost finished with their studies. *This is too late!* You have to think about your career and start planning and taking some action as soon as you start the program, or soon thereafter. You have to have a general idea about the vocation you are headed for so that you can take appropriate courses and pick up necessary skills; choose relevant or related dissertation research that will give you a background, an edge, or an early start; and meet the people who can help you. Importantly, be flexible; keep an open mind about the various opportunities available, and be willing to change direction if necessary.

♦ The keys to success are within yourself: Your biggest asset will be your *flexibility*; you will succeed if you take off the blinders and open your eyes to all possible career paths. Always maintain *optimism*; no one wants to work with someone who is negative. You need a strong cv with publications, and demonstrated *writing and presentation skills*. You must prove that you are a *team player.*

If you have the four things mentioned above then you are free to work in areas that are not considered "hot," and employers will be interested because you possess the very valuable people skills needed for success in today's changing scientific environment. (Ph.D., Chemistry, University of Rochester)

♦ People should be thinking about what kind of job they can get. Do they want to go into academics or do they want to go into industry? If industry, they have to start building a set of practical skills. They can't think that if they don't make it in academics, oh well, they'll just get a job in industry. Industry doesn't say, "Well, he's a smart guy, and that's enough." People have to have at least a set of skills that enable them to learn on the job. (Ph.D., Physics, University of Chicago)

How you can help yourself:

1. Take any courses in survival skills (job searching, interview skills, grantmanship, scientific writing, public speaking, teaching, etc.) that your university offers. Such training will be of benefit whether you are heading for academic or nonacademic employment (for example, Ph.D.s are notorious for entering industry with poor communication skills). Also, look up "Survival Skills for Graduate Students" on the internet (http://www.his.com/~graeme/employ.html). The excellent advice offered here is based on a workshop developed by Beth Fischer and Michael Zigmond. At the same web site, or more directly at http://www.nextwave.org, you can reach "Science's Next Wave," an interactive forum on traditional and nontraditional scientific careers and career-matters sponsored by *Science* magazine.
2. Consider taking a job at a small university or college. A somewhat different emphasis/preparation is required to procure a position at these types of institutions than at a large, research university. Interested students or postdocs should manage to obtain substantial amounts of teaching experience, and should take a range of courses in their field.

Small schools are great if you love teaching. You also may be able to do some research there, and you will not have to spend all your time writing grant proposals!

3. Develop highly useful skills (via courses or experience) that will help make you marketable. Become proficient in computer modeling, molecular biology techniques, etc. Ask practicing researchers where they think the future lies and, if the work interests you, do your dissertation research in a field that will likely be on the upswing. Try to use some techniques in your research that are needed in industry in case you decide to go in that direction (call up some industrial scientists and ask for advice).

◆ Scientists *can* be employable. There really are jobs out there. I had a pile of good job offers after getting my Ph.D., and my friends did too. None of us went to a top-notch school; none of us had great grades. Part of the problem is that a Ph.D. degree in science is an unbelievably specialized education; it often means finding some little corner of science that no one has explored, and becoming the best in the world at it. To be employable, you need to be very smart about which corner you choose—it must have some *possibility* of economic value. The *area around* your corner must have *immediate* economic value.

But how can students tell if an obscure area might have economic value? Most importantly, look at the funding sources of your prospective advisor. If no sources are private sector or task-related, short-term contracts, it is likely that there is little economic value to the work (even if it is scientifically wonderful). Other clues can be gained from papers published in the field; if they are all by academics, there probably aren't many job opportunities.

To find out about opportunities in the area around your little corner, look for recruitment ads in society publications, trade journals, big-city newspapers, and use them as feedback to yourself about your future employability. If you don't see ads for jobs that you would be immediately qualified for, you need to think seriously about the track you are on. Graduating grad students are another excellent way of determining your future employability. If some in your area are going into industry, it is a good sign that your field is economically viable. (Ph.D., Physics, University of Alabama)

4. Consider developing expertise in *interdisciplinary* fields. A student of chemistry who specializes in the biogeo-

chemistry of toxic metals may someday have to portray herself as either a biologist, geologist, or chemist in order to secure an available faculty position. A student of physics who deduces patterns from time-series data, and applies his techniques to study patterns of motion in foraging birds, may end up using his expertise at a biotech company analyzing the heartbeats of cardiac patients (the need for flexibility and versatility in industrial research was mentioned earlier). The harder it is for certain potential employers to classify you, the better your chances may be. There is also a need for people with backgrounds in both science *and* business, law, or government (for example, a Ph.D. or a master's in physics, and a M.B.A.), so you may want to think about going for an additional degree. Many Ph.D.s go on to medical school; besides the obvious advantage of being able to practice medicine, M.D./Ph.D.s make more money, find it easier to obtain research positions, and have additional funding areas available to them.

♦   I was a grad student in the laboratory of the Nobel Prize winner, Melvin Calvin. The Chemical Biodynamics laboratory contained a broad range of disciplines, with members specializing in biophysics to psychology, with experimental emphasis on biochemical to physical chemical approaches to understanding photosynthesis. This interdisciplinary environment had a profound effect on my career and I now work in 2–3 research areas including neurobiology and materials science, using techniques in transmission electron microscopy that I have subsequently developed but were first begun in Berkeley. (Ph.D., Chemistry, University of California, Berkeley)

5. Learn to work well in teams. Scientists are traditionally used to working alone, and often have trouble in industry where teamwork is the norm. Collaborate, join committees, or do other activities that will help you become more of a "people person."
6. See if you can do dissertation research that puts you in contact with future employers. For example, a geology student can do a project that involves hooking up with the United States Geological Survey, or a chemistry student can do research that requires some collaboration with a

pharmaceutical company. You will make important contacts, and develop the skills that the employers are looking for.

7. If you are interested, and if possible, take an internship or a summer position in industry; some departments will allow you to do this. You may find that you like industry better than academia. There is no teaching, no committee work, no constant grant writing; there is also more money (usually *much* more money), and a chance to see the direct benefits of your work. However, there is also a profit motive, an emphasis on collaborative research, no tenure, and possible restrictions on publishing (the secrecy issue) and what you can work on.

8. Consider doing a postdoc in an industrial lab. Although there is often a circumscribed area that you will be allowed to study, doing an industrial postdoc can be a very practical move, and can lead to a permanent job. Some companies (for example, DuPont and Genentech) allow postdocs to do basic as well as applied research. Inquire if you will be allowed to publish your work; a restriction on publishing will be a disadvantage if you want to return eventually to academia.

9. If you are interested in performing research for industry, or in working for a technical corporation in a managerial position where your scientific knowledge will be useful, it might serve you well to take some business courses, such as management, accounting, intellectual properties, etc. Ph.D.s with business skills are particularly attractive to employers (some business schools, such as the Johnson Graduate School of Management at Cornell University, offer special M.B.A. programs for people with advanced degrees in science). Also learn to network like crazy. Call successful people in industry, and ask for advice; ask for the names of others to call. It is a numbers game—the more people you speak to, the greater your chances of making contacts who can eventually help you to get a job. *Most jobs outside of academia are never advertised but are obtained through contacts.*

10. An ambitious student may be considering eventually starting his or her own small company, such as a biotech, computer software, or consulting company. This move

will be teeming with new and difficult challenges, such as raising money and making corporate ties. Keep asking for advice from those who have made it—you are going to need all the information that you can get. While the work is hard and the risks great, if you crave the excitement of being an entrepreneur, the rewards from starting your own company can be enormous.

As this chapter and the preceding one reveal, students struggling daily through their graduate programs cannot afford to get so absorbed in their scientific universe that they neglect worldly practicalities. Students must eventually define their goals and, whether they are heading for academic or nonacademic pursuits, must delineate and utilize those strategies that will lead to career success. Graduation day is not the time to start taking action.

17

The End Is in Sight:
Writing the Dissertation

Something strange happens to students in their last year as they struggle to finish up their final experiments. As they see the end approaching, as they perform techniques for the last few times (techniques they may have performed hundreds of times before), they grow more and more restless, more and more impatient. Even recalling what they have been through—the stress, the hurdles, the physical work—is exhausting. These students are weary of the whole graduate school thing, weary even of the projects that once held their interest. In short, they have reached the point where they can no longer stand being students. They picture what life will be like at the next, perhaps postdoctoral, stage of their careers; they seek a fresh start on exciting new research in some other lab, perhaps in some other part of the country. The work they are doing now is holding them back; every little delay at this point, such as the need to do another analysis, an experiment that has to be repeated, an error that has to be rectified, is irksome and intolerable. They have reached the breaking point. They want out.

Unfortunately, it is around this time and with this attitude that students are faced with the looming task of composing the dissertation. Now that experiments are being completed, that dreadful burden of writing down all that they have done and perceived is upon them.

Almost all students seem to fall apart to some extent during the writing of their dissertation. Part of the explanation lies with the general uneasiness and restlessness that we noted is typical at this stage. Also, "short-temperedness and irritability" is anticipated and even "allowed" at this point, and students are all too happy to take advantage of this emotional freedom. Yet surely the major explanation for the emotional difficulties experienced by students preparing the dissertation is the sheer enormity of the task that they face.

♦ It took me three months to write my dissertation and it was a stressful time. It is such a huge undertaking and it is *so* time consuming. (Ph.D., Chemistry, University of Rochester)

A student must compose a "book," 30 to 300 pages long, including tables and figures (math and physics dissertations tend to be on the low end of the page-count scale). Dissertations are written largely in the style of, and with the precision and depth of a journal article, and describe the sum of one's research efforts over many years. This is a daunting task (and one that takes months to a year to complete) that would intimidate even the best and most proficient of writers.

♦ Writing the dissertation was stressful and exhausting. In the end, I had to work very hard to put all other distractions aside and force myself to focus on completing the writing. But, I can say that after the fact, it was worth the experience. (Ph.D., Computer Science, University of Southern California)

♦ I felt burnt-out on the subject. It took tremendous self-discipline to push through since I was so tired of thinking about the topic and was planning to change fields in my postdoc. (Ph.D., Psychology, biopsychology research emphasis, University of California, Berkeley)

♦ Writing was definitely boring, because I knew it was not going to be published, and I was just going through the process. I just wanted to get it over with. From 6 a.m. to 8 p.m., I spent about 75 percent of my time writing and rewriting repeatedly, and doing analyses and graphs. It took 3–4 months; simultaneously I was finishing experiments and trying to get a postdoc position. (Ph.D., Physiology, Kansas State University)

♦ Twenty years ago, we did not use computers! Just typewriters, and if yours had the "self-correcting extra," life was heaven. The hardest part of the preparation of my thesis was satisfying the university librarian, whose job it was to have the margins perfectly set, no typos, no erasure marks, etc. If I made a typing error, sometimes I would have to type 3–5 pages to get things realigned again. As a testament to how difficult this task was, the woman who helped me type my thesis has been with me ever since and is now my wife. (Ph.D., Biological Sciences, University of California, Irvine)

If the student has been publishing his work all along, the writing of the dissertation should not be too much of a challenge, as much of the analysis, data representation, and writing has already been done and only needs to be transferred. Many departments allow three first-author journal articles, sandwiched between a general introduction and a general discussion and conclusion, to suffice as a dissertation.

◆ My dissertation was 3–4 research articles bound together. I enjoyed most the final discussion where I tried to build a larger, more synthetic theory. I learned to write *all day*. (Ph.D., Neuroscience, Dalhousie University)

Students who are not generating data worthy of publication as they progress through the years, are strongly urged to *write up each study as it is completed* anyhow. If they do not, and hold off writing until all experiments are done, the task before them is overwhelming.

◆ For me, writing the dissertation was not very stressful. I knew my dissertation would be long and have a rather extensive theoretical part, so I deliberately started writing before finishing all the experiments. I would write for a month or so to give my advisor another chapter to read, and then go back to the lab for a month. So the better part of a year elapsed between starting to write and finishing, but it was a very humane way to do it. I was very proud of my dissertation when I finished it, and although parts of it seem a little scientifically naive to me now, I still think it was perhaps the best thing I've ever written. Never again would I have the time and focus to pull so much material together to tell a story in a consistent voice from start to finish. (Ph.D., Chemistry, University of California, Berkeley)

To get an idea as to how these tomes are prepared, you should get copies of dissertations of prior students in the department, and skim through them (dissertations can be found in the university library or in the department office). You will note that a separate chapter is devoted to each delineated study, and that each study consists of one or more related experiments. Chapters are structured similarly to journal articles.

The general Introduction of the dissertation, the section devoted to reviewing the background literature and emphasizing the significance of the experiments undertaken in relation to what is already known, should not be difficult to prepare if you played your cards right. It has been previously mentioned that the easiest way to gather and sort out information for this section is to make use of what you have already written for your committee in the form of the dissertation proposal. Furthermore, if you have been clever throughout the years of graduate school, you might also have *purposely chosen topics relevant to your dissertation area for class papers, mock grant proposals, talks, or parts of the qualifying exam.* Thus, information for your dissertation, especially for its literature review, has been accumulating and you may have already written much of it down. Constructing the dissertation will involve piecing together these valuable strips of material. You should also have saved any posters that you presented so that text and figures (especially photographs) can be reused in relevant chapters.

A "Results" section (where the data are listed but not explained) and a "Discussion" section (where the meaning/significance of the results in regard to the literature are elaborated), may be included with each separate study. *Alternatively,* the results and conclusions from each study may be assembled together to form a general Results chapter and Discussion chapter. The Final Conclusions chapter includes the most important results, their significance, how they relate to the literature and the original hypotheses, and suggestions for future research. Composing this chapter is actually a rare opportunity. It is here that you can fashion your many studies into a connected whole and, drawing upon your results, can hypothesize a bit more broadly and freely than would be considered acceptable by a journal reviewer. This chance to be sweepingly creative and comprehensive will probably never occur again.

Procrastination is a very big issue for students writing the dissertation. It is extremely hard to sit down and write. The task is so huge that many do not even know how to start, or they have trouble pulling everything together. You will find that you will generate many ideas and explanations as you write—everything is not clear and coherent right from the beginning. Try not to carry compulsiveness to extremes. Your dissertation doesn't have to be absolutely perfect, totally comprehensive, and all-inclusive. Your committee will probably be the only ones to ever read it.

♦ I guess the biggest horror story imaginable would involve someone accidentally deleting the only copy of their entire dissertation on the computer. *Always have a hard copy, and a number of backup discs that you keep in safe and separated places.* A friend of mine was so afraid of losing her developing dissertation, that she kept one hard copy in a refrigerator in case there was a fire! (Graduate student, Earth and Planetary Sciences, Harvard University)

It is typical for a student to write one chapter of the dissertation, and then hand it over to her advisor for review and suggestions. After the student makes the changes recommended (or insisted upon), she writes the next chapter, and repeats the process. This continues, chapter after chapter. This is a difficult time for both student and advisor. For the student, it is painful to see her efforts buried in red pen markings; a very demanding advisor can easily add months to the writing time. For the advisor, the task is long and boring but, since the final product reflects on him, it must be done carefully and critically. It is sometimes necessary for the student to "bug" the advisor to hurry up and finish the dissertation reading. The quote below describes a student whose advisor took an inordinate amount of time to review the dissertation. It is an extreme and highly unusual case (even a bit funny), but to those who have been through the graduate school experience, it is certainly believable. Its lesson should extend beyond the stated situation. It should jolt students into realizing how unique the world of graduate school is, how truly on their own they are, and how they are going to have to look out for their own interests and pull their own strings in order to get what they need to succeed in their scientific careers.

♦ [I obtained my Ph.D. degree in Physics in 1958.] After years of intense studies, I finished my Ph.D. projects and presented the draft of the thesis to my advisor, Prof. X. I went to him after one month and asked, "Prof. X, did you read my thesis?" His answer was, "Sorry, I moved recently into a new house and there was no time." It was surely a reasonable explanation. However two months later, the same question was answered by "Sorry, my child was sick." He gave several other excuses every time I went to him. Then he moved to another University. Two years later, our Department Chairman asked me how my thesis work was progressing. When I told him that I wrote it years ago but Prof. X did not have the time to read it, he was shocked, and immediately proceeded to demand that Prof. X do something quickly. In this way, I finally obtained my Ph.D. degree after wasting more than two years. If my Department Chairman had not intervened on my behalf, I would still be without my Ph.D. degree after these 40 years. Now, being a professor myself, I understand that reading the Ph.D. theses of my students is often rather boring. However, because of my experience, I always manage to finish its reading within a month or so. (Ph.D., Physics, University of Rochester)

18

The Final Oral Exam (The Defense)

The final oral exam is the culmination of all that you have gone through since the day that you entered the graduate program. After years of taking tests, writing papers, making presentations, performing experiments, composing the dissertation, and jumping through numerous other hoops, it all boils down to this one-to-three-hour question-and-answer trial before the dissertation committee. And as momentous as it is, it is all over in an adrenaline-rush blur.

Departments differ in how they run their final exams—a few do not even have them. Many require that their students give a one-hour public presentation of their research right before the closed grilling session with the committee; to this open talk come the student's friends and relatives, lab members, and those who saw the notices about the talk that the department puts up on walls and bulletin boards, and thought the topic looked interesting.

♦ I was more stressed about the public talk than I was about the closed session of my defense. I was more worried about making a fool out of myself in front of my friends than in front of my committee. As soon as the talk was over I relaxed. (Ph.D., Chemistry, University of Rochester)

Some departments do not have their students give a public talk. Instead, students jump right into the oral exam. Officially, this is open to anyone, but only very rarely do people outside of the committee attend. This is the final chance for the committee to test the student's understanding of the work, find fault with the research and/or the writing, and examine presentation skills. The student is often first asked to step outside the room for a few minutes while the committee reviews the rules of the defense and the student's records. The student is then brought back inside, and proceeds (if there was no public talk) to give a fifteen-minute to one-hour review of his or her work, aided by slides or transparencies. Questions from the committee can interrupt the talk at any point, and they continue after the talk is over. The whole exam takes about one to three hours.

The defense is considered a formal affair. As such, students "dress up" for it as they did for the preliminary oral exam. Since research is performed in very casual attire, this may be the first time the student has dressed up in three or more years! Another popular tradition is to bring food for one's committee. In an attempt to appease the committee "gods," students bring pastries, fruit, bagels, or other goodies to the examination room for the committee to nibble on during the exam; the student, however, does not dare partake!

♦ I figured that if I scheduled my defense for a Friday afternoon, and brought *little* food, the committee would be extra eager to leave for the weekend, and wouldn't let the defense go on too long. And it worked! (Ph.D., Psychology, biopsychology research emphasis, University of California, Los Angeles)

Students who will soon be taking the final oral exam are stressed and pressed for time. To put them more at ease, friendly colleagues are quick to offer words of support and advice. Three comments in particular are repeated year after year to those preparing to defend, and *all ring of the truth*. So, when your time comes to take the big plunge, keep in mind:

1. "You know more about your dissertation topic than your committee does." For years, you have kept up with the literature, performed experiments, and thought long and hard about the issues. It has been your life. You are indeed more up to date and knowledgeable about your particular little corner of the universe than anyone else. Besides, truth be told, it is likely that some of your committee members have only *skimmed* your dissertation!
2. "It is a rubber stamp." That is, it (the defense) is only a formality, a done deal; if a date for the defense has been agreed

upon, it is likely that the committee has already decided that you have done enough work and exhibited enough logical thinking to be granted the degree.

♦ My final oral exam was a rubber stamp, although they did ask me very difficult and thoughtful questions. (Ph.D., Physics, Princeton University)

3. "Your advisor wouldn't let you take the exam unless he thought that you were ready for it." This is true. If you look bad, then your advisor looks bad, too.

There are some additional words of advice that should be mentioned concerning the exam and your preparation for it:

Make sure that you are in control of the situation. As ridiculous as this may sound, considering that you are the one forced to stand up there while that all-powerful committee takes pot shots at you, it is important that you maintain at least the façade of being in control. Your presentation should be done with an air of quiet confidence; do not exude bravado, but be steady, determined, mature, and self-assured. During your presentation and the committee's questioning, never let the situation get away from you. Act as if you are giving a seminar that is occasionally interrupted by questions from the audience: project a subtle demeanor of authority, of gently being in charge. Answer questions as if you enjoy doing so, as if you are merely transmitting information to an interested party; do not act as if you are being put on the spot or intimidated. Never mumble, hem or haw, or speak to the floor. Do not revert to a childish grimace or a defeated slouch. Do not appear apologetic. If you do not know the exact answer to a question (and are pretty sure the committee does not either), but you can make a good guess, do so. If you have no idea as to the answer, say that you do not know—your committee will not know the answers to all the questions either. Feel free to take a little time to think about your answers before speaking. After a question, gently direct attention back to your talk. If you disagree with a committee member, politely defend your position.

♦ A friend of mine told me about his experience at his preliminary oral. His advisor had complained that he had lost control of the group by appearing immature and flustered when challenged, and by just standing around awkwardly, waiting for someone to ask him a question or to tell him to continue with his talk. He improved for his final oral. I took his lesson with me to my own oral exam, and have been told that I presented myself quite well. (Ph.D., Structural Biology, Stanford University)

Anticipate questions that may be asked. We suggested previously that at your preliminary oral exam, you should think up in advance possible questions that your committee may ask you. The same advice goes for the final oral exam. Make sure that you reread your dissertation, piece by piece, before the exam; it has been many years since you performed some of the early experiments and you may easily have forgotten exactly how you did them, or all the reasons you did them for. Never leave anything open-ended. For example, if you wrote that your work ruled out that CCK plays a role in the development of the gut, and proposed that some other local but unnamed peptides must be involved, make sure that you can suggest what some of these other peptides may be if you are asked. Know how the assays and technologies you made use of work. Know the specialties of your committee members; they may ask you questions on topics in your research that they are familiar with. If individual members have expressed particular concerns or questions during past meetings, be able to address them. Make sure that you have answers to the "big" questions: What are the meanings of your results? How do they advance the field or relate to previous findings? Can you summarize your research and its relevance in a few sentences? What are the limitations or weaknesses of your techniques/study? If you had the chance to redo your experiments, knowing what you know now, would you do anything differently? What have you personally learned (skills, etc.) from doing your research? If you were to continue along this line of work, what other experiments would you run? Some of the questions may be difficult in order to challenge you and make you think, or just to shake you up. Ask your advisor to suggest some questions that he or she suspects may be asked. Hold a mock presentation in front of other grad students, and have them drill you. Finally, skim the latest journals in your field—some committee member may ask about a very recent finding that relates to your work.

♦ I knew the data better than anyone. What I didn't always know was how to put my data into a larger context. But it is this "big picture" that requires maturation and experience. The key to passing a dissertation is knowing what you know and what you don't know, and being able to steer the discussion towards your strength. (Ph.D., Psychobiology, University of California, Irvine)

♦ I dreaded my final oral, so the weeks just prior to my exam were very stressful—probably more stressful than the event itself, since it couldn't possibly be as traumatic as I imagined it would be! In fact, my oral presentation went smoothly. Some of the questions from my committee were difficult (I had to say I didn't know the answer), but overall my committee was reasonable. The day after my exam, I bungee-jumped in a symbolic release of all the tension I had felt those last few months. (Ph.D., Biology, University of Texas, Austin)

Every so often, the usually private world of graduate school makes news and is brought to the public's attention. In 1996, a graduate student defending his thesis before his committee at the San Diego State University in California pulled out a handgun and opened fire. Three of the professors were killed.

We assume that the problem behind this tragic event lay with the defendant—graduate students are not exempt from instability—and had little to do with the demands of the committee. Indeed, most committees actually turn out to be fair and just, and perhaps even kind. And most final orals turn out to be far less traumatic than expected. Many professors actually seem as if they are having fun and are enjoying themselves: they laugh and joke and kid around. If you are lucky, they will argue a point amongst themselves, seemingly forgetting that it is *your* defense that they are attending! You, too, will likely soon relax and begin to enjoy this unusual, once-in-a-lifetime situation. As this former student, now a professor, writes:

♦ The oral exams are always much more stressful than necessary. They loom as major stepping stones for the students, but the faculty have been through them many times, so they take less interest in them. The vague indifference can cause further anxiety for students, but they should keep reminding themselves that faculty have a lot of fish to fry, and they should not be offended by the faculty's hectic pace. Faculty do not have the time or interest to pick out a student to harass. (Ph.D., Psychobiology, University of California, Irvine)

Do not be surprised if the committee requests that you do some additional statistics, or that you rewrite some sections of the dissertation; minor requests are to be expected. Infrequently, a student is given a provisional pass, and is asked to perform some further experiments (a second defense is usually not required); if demands are excessive, the committee chairperson should stand up for the student. It is very rare for a student to downright fail the exam; if justified, both the student and the advisor/committee may be to blame. No one should be allowed to reach this final point if incompetent or not ready. If the student fails just the oral questions, but has an acceptable dissertation, a second defense may be demanded. If the student feels failure was unjustified, he can petition to have outside reviewers examine the situation.

♦ About two-thirds of my dissertation had already been published, so I felt reasonably confident and not very scared about going in. You know more about the stuff than anyone else, because you've lived with it for years. Also, your advisor shouldn't let you go up there unless you are ready. It's your thesis advisor's job to prepare you for the defense; we went over what had to be said and what we were going to present. When I walked into the room, they told me, "It's a good dissertation, and we intend to sign—start talking." I started talking. The first slide was the title of the dissertation, the second was a list of things I was going to talk about, and when I put on the third slide, they started asking me questions, and they asked me questions for the next two hours. And then we were done. I was surprised that they asked as many questions as they did. Towards the end, the questions were general like "Did you think of doing this?" Well, yeah I did, but I wanted to graduate. They asked a lot of intelligent questions. I was exhausted when I finished. (Ph.D., Physics, Brown University)

♦ I wasn't scared. I knew I could handle any question—I had been to enough seminars and professional meetings, and I saw how people could dance around pointed questions with non-pointed answers. The exam went very smoothly. I gave a one-hour presentation, and the committee asked me questions for about three-quarters of an hour. Some of the professors did research completely outside my area, and they asked questions because they were interested, so the questions were easy. Afterwards, my advisor and I went to a taco place for food and beer. (Ph.D., Psychology, biopsychology research emphasis, Vanderbilt University)

When the committee has finished its questioning, you will be asked to leave the room and wait out in the hall. This is when the committee decides if it should or should not pass you. This will be an awkward time for you. As you stand there alone, time passes very slowly, and imagination runs wild.

> ♦   One piece of advice for students waiting for the results of oral exams is to remember that most faculty members like to talk and that exams can be social events. Therefore a long delay may have nothing to do with you but can be due to extraneous conversations. (Ph.D., Biochemistry, Stanford University School of Medicine)

When the door finally opens, someone will usher you inside. Most probably you will be met by committee members with smiling faces and extended hands, greeting you with "Congratulations, Dr." You smile with relief. The committee members all sign the signature page of your dissertation and, soon after, disperse. Your lab may take you out for lunch to celebrate. Then everyone will go back to doing what they were doing. Like so many other stages and aspects of graduate school, this closing chapter is characterized by vagueness and understatement. And you progress alone, a graduating class of one.

All in all, after so many years, it is anticlimactic.

Handing in the Dissertation

Passing the final oral exam makes you unofficially a Ph.D. *Officially* you are left hanging in space. You do not really get your degree until the day of the graduation ceremony. *You are not really finished with all your degree requirements* until you perform the additional work or/and make the corrections to the dissertation that your committee has insisted upon at the final oral. You must also contend with the frustrating details of formatting the dissertation according to official, university specifications. Every dissertation (a copy of which goes onto the shelves of the university library) is stored on microfilm, and must meet strict, standardized guidelines regarding left and right margins, top and bottom borders, layout of pictures and tables, numbering of pages, wording of introductory pages, type/weight of paper used, etc. You go a bit crazy trying to comply with all the requirements.

As you work, the end being here, you sense that you are starting to exit the student mold. Your opinions about graduate training are shifting and softening. You glimpse the more extensive change in attitude towards the educational process that occurs gradually as one gets further and further away from graduate school. Things begin to make sense, the pieces fall together, the role of the advisor becomes more apparent, and a clearer, more positive perspective develops. For those who remain in academia, this culminates when the former student becomes a professor himself and is training students of his own.

For now, though, the spelling errors must be corrected, the margins must be reduced, the font must be as specified, and the copyright page must be added. The last step is to turn in the dissertation—it is not until you hand over your dissertation and have its carefully prepared format officially approved by the proper administrator that you can "truly" consider yourself to have completed graduate school. One fine day, this small, yet significant, final act occurs quite unceremoniously, and graduate school is left behind.

♦ It was nerve wracking making sure that all the dissertation regulations were met: the margins, the borders around pictures, 1" here, 3/4" there, the format of the copyright page, the words on the title page, the lines on the signature page, the Table of Contents, etc., etc.—all had to be precisely as specified. I really wasn't sure that I had it all right. I had heard that the woman in the main library that you turn the dissertation in to, and who checks everything, was very strict. Everyone at this stage was "scared" of her. I remember thinking when I saw her that she looked younger and nicer than I had expected. The room was scattered with dissertations of all thicknesses—on desks, on shelves, on the floor. I stood there while she went through my dissertation, page by page. After all those years, it was coming down to this. Eventually she looked up, smiled, and said in a sweet voice, "Everything seems fine—congratulations." Just like that, with those few words, it was all over. No ceremony, no trumpets. I walked outside in a daze. I stood and looked at the campus for a long, long time. I had finished. (Ph.D., Psychology, biopsychology research emphasis, University of California, Los Angeles)

References and Additional Reading

Chapter 1

Barnes, Gregory A. *The International Student's Guide to the American University*. Lincolnwood, Ill.: National Textbook Company, 1991. This is a clear and easy-to-read little book with practical information for foreign students considering graduate education in the U.S. Includes a useful bibliography.

Hamel, April V., with Mary M. Heiberger, and Julia M. Vick. *The Graduate School Funding Handbook*. Philadelphia: University of Pennsylvania Press, 1994.

Moore, Richard W. *Winning the Ph.D. Game*. New York: Dodd, Mead & Company, 1985. An informative and even humorous book, with good chapters on choosing a university and getting admitted.

Peters, Robert L. *Getting What You Came For: The Smart Student's Guide to Earning a Master's or a Ph.D.* New York: The Noonday Press/Farrar, Straus and Giroux, 1992. Excellent book on winning strategies for graduate school, although not specifically for science majors. Tips on choosing a school, getting your foot in the door, and picking an advisor. Useful information for both master's and Ph.D. students.

Chapter 2

Committee on Science, Engineering, and Public Policy, National Academy of Sciences. *Adviser, Teacher, Role Model, Friend: On Being a Mentor to Students in Science and Engineering*. Washington, DC: National Academy Press, 1997. Of interest to faculty and students. Available on the internet at http://www.nap.edu/readingroom/books/mentor.

Chapter 6

Beynon, Robert J. *Postgraduate Study in the Biological Sciences*. London: Cambridge University Press, 1993. Very informative for learning about scientific databases and the maintenance of bibliographic databases.

Smith, Robert V. *Graduate Research: A Guide for Students in the Sciences.* New York: Plenum Press, 1990. Offers excellent tips for searching the literature, keeping up with the literature, and filing reprints and photocopies of articles in an organized, easily accessible fashion. Principles and methodology of research. Ethics of science.

Chapter 9

Madsen, David. *Successful Dissertations and Theses.* San Francisco, CA: Jossey-Bass, 1988.
Peters, Robert L. *Getting What You Came For: The Smart Student's Guide to Earning a Master's or a Ph.D.* New York: The Noonday Press/Farrar, Straus and Giroux, 1992.

Chapter 11

Committee on Science, Engineering, and Public Policy, National Academy of Sciences. *On Being a Scientist: Responsible Conduct in Research.* Washington, DC: National Academy Press, 1995. Available on the internet at http://www2.nas.edu/cosepup.

Chapter 12

Bowen, William G., and Neil L. Rudenstine. *In Pursuit of the Ph.D.* Princeton, NJ: Princeton University Press, 1992.

Chapter 13

Barnes, Gregory A. *The International Student's Guide to the American University.* Lincolnwood, IL: National Textbook Company, 1991.
Normile, Dennis and June Kinoshita. "Science in Japan: Competition on Campus." *Science,* vol. 274, 1996, pp. 43–56.
Peters, Robert L. *Getting What You Came For: The Smart Student's Guide to Earning a Master's or a Ph.D.* New York: The Noonday Press/Farrar, Straus and Giroux, 1992. Reviews common problems experienced by foreign students.

Chapter 14

Booth, Vernon. *Communicating in Science: Writing and Speaking.* New York: Cambridge University Press, 1985.
Booth, Vernon. *Communicating in Science: Writing a Scientific Paper and Speaking at Scientific Meetings,* 2nd edition. New York: Cambridge University Press, 1993.

Day, Robert A. *How to Write and Publish a Scientific Paper,* 4th edition. Phoenix, AZ: Oryx Press, 1994.

Feibelman, Peter J. *A Ph.D. Is Not Enough: A Guide to Survival in Science.* New York: Addison-Wesley Publishing Company, 1993.

Matthews, Janice R., John M. Bowen, and Robert W. Matthews. *Successful Scientific Writing: A Step-by-Step Guide for the Biological and Medical Sciences.* New York: Cambridge University Press, 1996.

Smith, Robert V. *Graduate Research: A Guide for Students in the Sciences.* New York: Plenum Press, 1990.

Chapter 15

Beynon, Robert J. *Postgraduate Study in the Biological Sciences.* London: Cambridge University Press, 1993. Helpful for many technical matters, such as using computers to generate graphs and tables, designing slides, transparencies, and posters. Also, good pointers on communicating science, filing notes, maintaining bibliographic databases, and other matters of interest to researchers.

Booth, Vernon. *Communicating in Science: Writing and Speaking.* New York: Cambridge University Press, 1985.

Booth, Vernon. Communicating in Science: *Writing a Scientific Paper and Speaking at Scientific Meetings,* 2nd edition. New York: Cambridge University Press, 1993.

Braben, Donald. *To Be a Scientist.* New York: Oxford University Press, 1994.

Committee on Science, Engineering, and Public Policy, National Academy of Sciences. *Adviser, Teacher, Role Model, Friend: On Being a Mentor to Students in Science and Engineering.* Washington, DC: National Academy Press, 1997. While intended for faculty advisors, this publication lists excellent resources (regarding gender and minority issues, ethics, communication skills, time management, job-hunting, etc.) of interest to students and postdocs. Available on the internet at http://www.nap.edu/readingroom/books/mentor.

Feibelman, Peter J. *A Ph.D. Is Not Enough: A Guide to Survival in Science.* New York: Addison-Wesley Publishing Company, 1993. Includes practical tips on grant writing and public speaking.

Medawar, Peter, B. *Advice to a Young Scientist.* New York: Harper & Row, 1979.

Moffat, Anne Simon. "Grantsmanship: What Makes Proposals Work?", *Science,* vol. 265, 1995, pp. 1921–22.

Peters, Robert L. *Getting What You Came For: The Smart Student's Guide to Earning a Master's or a Ph.D.* New York: The Noonday Press/Farrar, Straus and Giroux, 1992. Great for tips on self-management and self-organization. Includes pointers on oral presentations.

Phillips, E.M. and Pugh, D.S. *How to Get a Ph.D.* Philadelphia, PA: Open University Press, 1994. Problems students encounter.

Simmonds, Doig and Linda Reynolds. *Computer Presentation of Data in Science*. Dordrecht, The Netherlands: Kluwer Academic Publishers, 1988.

Smith, Robert V. *Graduate Research: A Guide for Students in the Sciences*. New York: Plenum Press, 1990. Includes advice on oral presentations.

Tufte, Edward R. *Visual Display of Quantitative Information*. Cheshire, CT: Graphics Press, 1983.

Chapter 16

The following references elaborate on the disturbing employment situation in science today; personal strategies and/or changes in graduate education that can mitigate the predicament facing young scientists are discussed.

Ausubel, Jesse H. "Malthus and Graduate Students: Checks on Burgeoning Ranks of Ph.D.s," *The Scientist*, Feb. 5, 1996, p. 11.

Committee on Science, Engineering, and Public Policy; National Academy of Sciences; National Academy of Engineering, Institute of Medicine. *Careers in Science and Engineering: A Student Planning Guide to Grad School and Beyond*. Washington, DC: National Academy Press, 1996. Available on the internet at http://www2.nas.edu/cosepup.

Feibelman, Peter J. *A Ph.D. Is Not Enough: A Guide to Survival in Science*. New York: Addison-Wesley Publishing Company, 1993. Practical advice on securing a job (postdoctoral or permanent) in academic and nonacademic research. Includes interview strategies.

Fiske, Peter S. *To Boldly Go. A Practical Career Guide for Scientists*. Washington, DC: American Geophysical Union, 1996.

Gibbons, Ann. "Innovations on Campus." *Science*, vol. 266, 1994, pp. 844–51.

Griffiths, Phillip A. "Reshaping Graduate Education," *Issues in Science and Technology*, Summer 1995, pp. 74–9.

Heiberger, Mary M. and Julia M. Vick. *The Academic Job Search Handbook*. Philadelphia: University of Pennsylvania Press, 1992.

Holden, Constance, Peter Radetsky, and Anne Simon Moffat. "Science Careers: Playing to Win." *Science*, vol. 265, 1994, pp. 1905–10, 1921–22, 1931–32.

Holden, Constance, Anne Simon Moffat, Jocelyn Kaiser, Paul Selvin, and Karen Celia Fox. "Careers '95: The Future of the Ph.D." *Science*, vol. 270, 1995, pp. 121–42.

Lanks, Karl W. *Academic Environment. A Handbook for Evaluating Employment Opportunities in Science*, 2nd ed., Philadelphia: Taylor and Francis, 1996.

Massy, William F. and Charles A. Goldman. *The Production and Utilization of Science and Engineering Doctorates in the United States*.

Stanford, CA: Stanford Institute for Higher Education Research, 1995.

Moore, Richard W. *Winning the Ph.D. Game.* New York: Dodd, Mead & Company, 1985. Good information on getting jobs inside and especially outside of academia.

Research-Doctorate Programs in the United States. Washington, DC: National Academy of Sciences, 1995.

Committee on Science, Engineering, and Public Policy, National Academy of Sciences, National Academy of Engineering, Institute of Medicine. *Reshaping the Graduate Education of Scientists and Engineers.* Washington, DC: National Academy of Sciences, 1995. Available on the internet at http://www2.nas.edu/cosepup.

National Science Board. *Science and Engineering Indicators.* Washington, DC: Government Printing Office, 1993.

Peters, Robert L. *Getting What You Came For: The Smart Student's Guide to Earning a Master's or a Ph.D.* New York: The Noonday Press/Farrar, Straus and Giroux, 1992. Includes helpful advice for students thinking about employment inside or outside of academia.

Solmon, Lewis C., Laura Kent, Nancy L. Ochsner, and Margo-Lea Hurwicz. *Underemployed Ph.D.s.* Lexington, MA: Lexington Books, 1981.

Smith III, T. P. and Tsang, J. C. "Graduate Education and Research for Economic Growth," *Science*, vol. 270, 1995, pp. 48–49.

Tobias, Sheila, Daryl E. Chubin and Kevin Aylesworth. *Rethinking Science as a Career: Perceptions and Realities in the Physical Sciences.* Tucson, AZ: Research Corporation, 1995.

Chapter 17

Peters, Robert L. *Getting What You Came For: The Smart Student's Guide to Earning a Master's or a Ph.D.* New York: The Noonday Press/Farrar, Straus and Giroux, 1992. Includes helpful solutions to writers block. Discusses the structure of the dissertation.

Index

Admissions, 6–7
  application for, 6
  Admissions Committee, 8
  criteria for, 9
  deadlines, 7,8
  essay for, 10–11
  of foreign students, 16–19
  GRE (Graduate Record
    Exam), 9
  interview, 11–12
  letters of recommendation, 9
  the process, 8
  selecting schools, 6,7
  transcripts, 10
  research prior to, 5
  visits, 13
Advisor
  absence of, 50–54
  changing mentor, 34
  funding of, 29
  lab rotation with, 21
  meeting with, 60–61
  selection of, 20–34, 83

Careers, 3, 161–63, 175–85
  in industry, 181, 184, 185
  job market, 175–80
  in teaching, 4, 181–82

interdisciplinary, 183
Committee. *See* Dissertation
Competition, 128–30
Comprehensive exam. *See*
    Qualifying exam
Course work, 35, 41
  classes, 41–44
  grades, 43, 44
  journal clubs, 44–45, 59, 61
  seminars, 46–49, 61, 65
Criticism, 65–67

Degrees
  Masters degree, 4, 14, 140,
    178
  M.D. degree
    combined M.D./Ph.D., 4
    foreign students, 18
    unsuccessful applicant, 14
  Ph.D., 4
Dissertation
  choosing dissertation re-
    search, 70–80
  committee, 90, 93
  final oral defense, 40, 46,
    192–97
  handing it in, 198–99
  preliminary oral exam, 90–98

proposal, 38, 63, 186–91
  writing, 39, 63, 185–91
Dropping out, 139–40

Field work, 103–5
Financial aid/stipends, 14–15,
    180. *See also* Teaching
    Assistant
Foreign students, 16–19,
    141–49
  as applicants, 16
  TOEFL (Test of English as a
    Foreign Language), 18, 45
  culture shock, 146–47

Graduate student
  competition, 128–30
  luck in research, 130–31
  social life, 116–18
Grants, 63
  of advisor, 29
  writing proposals, 165, 182

International students. *See*
    Foreign students

Jobs. *See* careers
Journals
  reading the literature, 61–62
  writing for, 63, 150–59
Journal clubs, 44–45, 61
  stress in, 133

Laboratory
  choice of, 20–34
  competition in, 128
  etiquette in, 122–28
  life in, 101–18
  meetings of, 45, 46, 58–60
  notebooks for, 119–22
  rotation through, 21, 28
  size, 32
Meetings
  with advisor, 60–61

and networking, 86–88
  poster presentations at, 62,
    81
  scientific, 62, 81, 82, 86, 87

Networking, 48, 52, 63, 163,
    169
  in choice of advisor, 83
  in department, 84–85
  at meetings, 86–89

Postdoctoral, 169–72, 184
  of potential advisor, 30
  in industry, 171
  publications during, 172
  research, 52
  as teachers, 57
Posters. *See* Meetings
Publications
  importance of, 74, 109
  writing journal articles,
    150–59

Qualifying examination, 35–39
  preliminary oral, 39
  criticisms of, 67

Research, 39
  choosing dissertation re-
    search, 70–80
  during lab rotation, 21
  ethics, 122
  lab notebooks, 119–22
  learning how, 55–60
  mistakes in, 64–65
  publishing, 109
  routine of, 106
  as an undergraduate, 5

Seminars, 61, 166, 167, 85. *See
    also* Course work
  and Networking
Stress, 130–39

and coping strategies, 116–18
and dissertation defense, 193
of foreign students, 144, 149
and oral exam, 94, 132, 134,
   136
and picking an advisor, 135

Teaching assistant, 13, 15, 16,
   101, 180
Thesis. *See* dissertation
TOEFL (Test of English as a
   Foreign Language), 18, 45.
   See also Foreign students